FROM THE VIEWPOINT OF THE LVOV-WARSAW SCHOOL

POZNAŃ STUDIES
IN THE PHILOSOPHY OF THE SCIENCES AND THE HUMANITIES

VOLUME 78

This book has been partly sponsored by Adam Mickiewicz University

The address: prof. L. Nowak, Cybulskiego 13, 60-247 Poznań, Poland.
Fax: (061) 8477-079 or (061) 8471-555
E-mail: epistemo@main.amu.edu.pl

FROM THE VIEWPOINT OF THE LVOV-WARSAW SCHOOL

Jacek Juliusz Jadacki

Rodopi

Amsterdam - New York, NY 2003

The paper on which this book is printed meets the requirements of "ISO 9706:1994, Information and documentation - Paper for documents - Requirements for permanence".

ISSN 0303-8157
ISBN: 90-420-0904-7

Printed in The Netherlands

CONTENTS

PREFACE

One of the most striking peculiarities of the Lvov-Warsaw School is the fact that its representatives have taken genuinely collective inquiries. I am not an exception. Most of the studies inserted in this volume are results of such a scholarly collaboration.

The study "On What Seems not to Be" is a reflection of my discussions with Professor Marian Przełęcki, my master in formal logic. The impulse to deal with problems exposed in the study "On the Controversy about Universals" was given by Professor Jerzy Pelc, my master in logical semiotics. The long co-operation with Professor Zdzisław Augustynek, my late master in ontology, was the main source of the ideas expressed in the study "On Forms of Objects". The point of departure of the study "On Good, Necessity and Sufficiency" was one of the chapters of a beautiful and clever book *Język i wartości* [Language and Values] written by Professor Jadwiga Puzynina and a long discussion with the author. The study "On Essence" relates to my works connected with publishing — together with my former pupil and present scrupulous collaborator, Dr. Kordula Świętorzecka — the volume *Wiedza i wiara* [Knowledge and Faith], which contains collected papers of Rev. Jan Salamucha; had he not been murdered by Nazis during the last war, I am sure I would be his close follower in metaphysics. The genesis of the study "On the Definition and Criteria of Truth" reaches as far back as the times of a teamwork contribution of my Department to *Encyclopedic Dictionary of Semiotics* but a decisive inspirer was Professor Peter Thomas Geach, whose lectures, which I attended at Warsaw University, opened my eyes to new prospects in this sphere. Many difficulties concerning the questions analysed in the study "On Linguistic Categories" were indicated to me by my students, first of all by Dr. Mieszko Tałasiewicz. The study "On Questions" would have never come into being without the kind inspiration of Professor Roman Stanisław Ingarden, the son of Roman Ingarden, the philosopher; it owes its final shape to Aleksandra Białecka, my promising young student. Professor Andrzej Bogusławski, whom I regarded, at first, only as an example of a true linguist and now of a real scientist in general, provoked the considerations whose results are inserted into the study "On Semantic Functions of Conditionals". The study "On Nonsense" arises out of the contacts with Professor Klemens Szaniawski, my master in methodology. My interest in pragmatics, manifested in the study "On Misunderstanding about Understanding", was challenged by Professor Barbara Stanosz, who influenced me strongly in the earlier period of my university career. The method of analysing mental phenomena in the study "On Definition, Explication, and Paraphrase" I borrowed from Professor Jerzy Perzanowski, whose penetrating mentality takes a strong hold upon my views not only in the area of philosophy. I trashed out many doubts of the study "On Reasoning" with my another pupil, Dr. Joanna Odrowąż-Sypniewska. The starting point of the content of the study "On Simplicity" were results obtained by my old well-tried friend, Professor Witold Strawiński. It was Professor Kwiryna Handke who has persuaded me to speak about... silence; the study "On Silence" would have been much more imperfect had I not had the opportunity for discussion with the participants of her conference and with Professor Zdzisław Najder in particular. The study "Science and Creation", which closes this volume, is a version of the inaugural lecture at the Warsaw Academy of Music, which I gave in 1999 being invited by a distinguished composer — and my old friend — Professor Stanisław Moryto, *Rector Magnificus* of our common *Alma Mater*.

I am not able to indicate all these crucial topics which I was able to rethink and reformulate thanks to suggestions of my honorable colleagues, Professor Piotr Brykczyński and Professor Jan Woleński, as well as my talented pupils: Dr. Tomasz Bigaj and Dr. Anna Wójtowicz.

My heartfelt thanks go to all those, who are not, however, responsible for the final shapes of thoughts which they may recognise in this volume as only pale shadows of their own ideas.

* * *

Portions of some studies collected in the volume have been published before. I wish to express my gratitude to the editors and publishers for permission to include them here in a rather different form. The sources of the originals of these portions are as follows:

The Conceptual System of the Lvov-Warsaw School [1996]. *Axiomathes* **9** (3), 325—333;

Being and Existence. On the Being of What Seems not to Be [1981]. *Dialectics and Humanism* **8** (4), 131—139;

Controversy about Universals [1986]. In: Th.A. Sebeok (ed.), *Encyclopedic Dictionary of Semiotics*, vol. 1, pp. 1138—1141. Berlin: Mouton de Gruyter;

On Forms of Objects [1999]. In: L. Albertazzi (ed.), *Shapes of Forms*, pp. 341—359. Dordrecht: Kluwer Academic Publishers,

Truth: Its Definition and Criteria [1990]. *Dialectics and Humanism* **17** (2), 193—212;

Troubles with Categorial Interpretation of Natural Language [1997]. In: R. Murawski and J. Pogonowski (eds.), *Euphony and Logos. Poznań Studies in the Philosophy of the Sciences and the Humanities*, vol. 57, pp. 327—343. Amsterdam: Rodopi,

On Semiotic Functions of Conditionals [1988]. In: M. Herzfeld and L. Melazzo (eds.), *Semiotic Theory and Practice*, pp. 469—481. Berlin: Walter de Gruyter & Co.;

Misunderstanding about Understanding [1994]. *Acta Universitatis Lodziensis. Folia Philosophica* **9**, 33—41;

Definition, Explication, and Paraphrase in the Ajdukiewiczian Tradition [1995]. In: V. Sinisi and J. Woleński (eds.), *The Heritage of Kazimierz Ajdukiewicz. Poznań Studies in the Philosophy of the Sciences and the Humanities*, vol. 40, pp. 139—152. Amsterdam: Rodopi.

* * *

I find it difficult to express in adequately ponderable words my great thanks for inestimable assistance of Dr. Katarzyna Paprzycka and Dr. Joanna Odrowąż-Sypniewska, who, in fact, were the editors of this book: it owes its final form in all essential respects to their capacity and care.

Jacek Juliusz Jadacki

INTRODUCTION: PHILOSOPHY AND PRECISION

1.

Kazimierz Twardowski wrote in 1919 in his famous paper "On Clear and Unclear Philosophical Style":

> Human language [...] is not just a system of conventional signs and above all it is not merely an external manifestation of thought, but is also that instrument of thought which makes abstract thinking possible for us; when thinking, we think in words — that is, *via* language [Twardowski 1919—1920, p. 258].

This metaphor of language as a tool becomes intelligible in the light of two distinctions made in 1934 by Twardowski's great disciples: Tadeusz Kotarbiński and Kazimierz Ajdukiewicz. Kotarbiński — analysing the notion of act — distinguished i.e. its author, its tool, its material, and its product (i.e. its intended result) [Kotarbiński 1934]. Ajdukiewicz distinguished among the acts of thinking the articulated judgments (i.e. judgments expressible in sentences), and among the latter the verbal judgments (i.e. judgments containing acts of assertion and presentations of a respective sentence) [Ajdukiewicz 1934a, pp. 36—37].

Appealing to these distinctions, we can state precisely the positive part of Twardowski's thesis saying that in verbal thinking language plays the part of its tool as well as its material. The role of language in such a situation is analogous to the role of a piece of chalk during the act of writing on the blackboard.

The sentence:

(i) I claim that *p*

is an external *expression* of my verbal act of judgement, and the sentence:

(ii) *p*

is a *tool-material* of this act of judgement.

It is obvious that the logical analysis of the philosophical language is necessary if we want to work according to the standards of the Lvov-Warsaw School.

2.

In his well-known works "Sprache und Sinn" [Ajdukiewicz 1934a] and "Das Weltbild und die Begriffsapparatur" [Ajdukiewicz 1934b], Ajdukiewicz introduced the notion of *conceptual apparatus* or the class of all senses ascribed to expressions of a certain (consistent and closed) language. He contrasted this notion with the notion of *language*, on the one hand, and the notion of *world-image* or the class of all sentences, on the other. The latter can be constructed with the aid of notions belonging to a given conceptual apparatus and it should be accepted in the face of experiential data.

I understand the aim of analysing the philosophical language as reconstructing the conceptual apparatus of this language. I am aware of two tendencies here. The first one is a tendency initiated by Twardowski himself, namely the tendency towards an essentially enriched list of conceptual items. "Essentially enriched" means that new concepts are not fully defined in terms of old conceptual items only. The second one is a tendency connected mainly with Kotarbiński, namely the tendency towards reducing this list.

There is a kind of common background for these two tendencies. What I have in mind is a certain version of Occam's razor, i.e., the old scholastic dictum: *entia non sunt multiplicanda praeter necessitatem*. The relevant version goes as follows: before introducing new notions we should examine whether we could do without them. According to such a rule of conceptual optimisation, all concepts belonging to the domain of the relation of being-nothing-but should be eliminated from the philosophical conceptual apparatus.

I think that the rule of conceptual optimisation together with the rule of conceptual precision are the fundamental methodological rules scrupulously observed in the Lvov-Warsaw School.

Conforming to this rule in practice, we are obliged to find such a known relation to a known object which determines the notions under consideration. If it is impossible to construct a satisfactory product and to select suitable domains or counterdomains of former relations, we can and should find either new relations or a set of new objects that satisfy these conditions.

3.

The rule of conceptual precision as a current rule in the Lvov-Warsaw School was analysed in detail by Marian Przełęcki [1998], one of my teachers. I would like to stress only two aspects of this rule.

From the viewpoint of the Lvov-Warsaw School, the philosophical language should be determined precisely, in particular with regard to its semantic and pragmatic functions.

We can find many beautiful archaic metaphors in Kotarbiński's works. Tatarkiewicz's writings are distinguished by their elegant expressiveness. That is really true! But the relevant semantic function of the philosophical language consists in the fact that it carries merely literal senses. And only intellectual expression (i.e. the expression of beliefs, and not, for example, emotions or ecstasy) is its relevant pragmatic function.

Kotarbiński's metaphors are easily understood against the background of his literal considerations. Tatarkiewicz's expressiveness does not cover intellectual contents of his language.

In the light of these facts, when two outstanding representatives of the School, Tadeusz Czeżowski and Przełęcki claim that metaphorical utterances serve as expressions of our experience of a mysteriousness of the world or a sense of human life, and when they declare that such an experience cannot be, in principle, literally verbalised, their liberalism on this point should be treated as a serious breach of the methodological viewpoint of their (as well as my) School.

4.

For many philosophers, the ideal situation is when the elements of the conceptual apparatus of our philosophy match the elements of the «philosophical» world and *vice versa*. Thus, the second tendency mentioned above is accompanied by the postulate (present in Kotarbiński's writings) to regard the supposed objects as corresponding to «pared»[1] parts of language as hypostases required by ontological reduction.

This situation can be described by means of the distinction made by Jan Łukasiewicz in his "Analiza i konstrukcja pojęcia przyczyny" [Analysis and Construction of the Notion of Cause], published in 1906. He identified notions with abstract objects and distinguished among them two kinds: «ready-made» notions and notions that had to be constructed. According to Łukasiewicz, the latter can be either ideal or real.

> To construct an abstract object means to search out some properties, to take into consideration the problem of which of them can be interconnected, and which of them must be removed, and to obtain in such a way a set of properties, linked together with relations" [Łukasiewicz 1906, p. 13].

Among abstracts ideal notions are "such abstract objects which are constructed by our mind regardless of whether they refer to something in reality or not, i.e. do not aim to cover concrete objects". Real notions, on the contrary, are "such abstract objects, which are constructed by our mind on account of reality, i.e. which do aim at covering some concrete objects" [Łukasiewicz 1906, p. 14].

It seems to me that a characteristic feature of the viewpoint of the Lvov-Warsaw School is that philosophical notions should be created with the intention of being real notions.

5.

Andrzej Grzegorczyk wrote in 1993 in his book *Życie jako wyzwanie* [Life as a Challenge]:

> I present a manner of philosophising which is very primitive [...]. I would like to call it *proto-philosophy*. According to this manner making precise some relatively simple [...] notions, and formulating a simple [...] general hypothesis with the aid of them, is considered a cognitive achievement [Grzegorczyk 1993, p. 5].

Grzegorczyk is regarded as a representative of the Lvov-Warsaw School and he regards himself as its member. This fact is confirmed, e.g. by his initial ontological declaration, determining, as he writes, "the general

[1] Quotation marks of the form: «...» — serve to metaphorise a given expression (*passim*).

conceptual structure applied [...] to the description of the world" [Grzegorczyk 1993, p. 27]. He declares also that his book "is not a summary of its tradition but is a continuation of this tradition" [Grzegorczyk 1993, p. 6].

Deriving from the same tradition, I set both goals before myself in my philosophical activity: to summarise as well as to continue.[2] In many of my papers I try to make a clear arrangement of the heritage of the Lvov-Warsaw School as well as of the Polish analytical philosophy in general.[3]

My present book is my own outlook on matters of philosophy. I hope it is CLEARER now what is meant by saying that this book consists of PROTO-THEORETICAL STUDIES made FROM THE VIEWPOINT OF THE LVOV-WARSAW SCHOOL.

---

[2] One of symptoms of this continuation is that I consequently use here Polish notation.

[3] Cf., for example: On the Sources of Contemporary Polish Logic [1980]. *Dialectics and Humanism* **7** (4), 163—183; On Leon Chwistek's Semiotic Views [1984]. In: J. Pelc *et al.* (eds.), *Sign, System and Function*, pp. 77—87. Berlin: Mouton Publishers; Lwów-Warsaw School [1986]. In: Th.A. Sebeok (ed.), *Encyclopedic Dictionary of Semiotics*, vol. 1, pp. 478—480. Berlin: Mouton de Gruyter; On Roman Ingarden's Semiotic Views: A Contribution to the History of Polish Semiotics [1989]. *Analecta Husserliana* **27**, pp. 523—540; The Metaphysical Basis of Kazimierz Twardowski's Descriptive Semiotics [1992]. In: J. Paśniczek (ed.), *Theories of Objects: Meinong and Twardowski*, pp. 57—74. Lublin: Wydawnictwo Uniwersytetu M. C.-Skłodowskiej; [with W. Jadacka], [La philosophie polonaise] [1992]. Some scores of items in: A. Jacob (ed.), *L'encyclopédie philosophique universelle*, vol. 3. Paris: Presses Universitaires de France; Kazimierz Twardowski's Descriptive Semiotics [1993]. In: F. Coniglione, R. Poli and J. Woleński (eds.), *Polish Scientific Philosophy: the Lvov-Warsaw School. Poznań Studies in the Philosophy of the Sciences and the Humanities*, vol. 28, pp. 191—206. Amsterdam: Rodopi; [with M. Przełęcki] [La philosophie en] Pologne [1993]. In: R. Klibansky and D. Pears (eds.), *La philosophie en Europe*, pp. 323—347. Paris: Éditions Gallimard — UNESCO; Warsaw: The Rise and Decline of Modern Scientific Philosophy in the Capital City of Poland [1994], *Axiomathes* vol. **5** (2—3), 225—241; [reprinted in:] R. Poli (ed.), "*In itinere.*" *European Cities and the Birth of Modern Scientific Philosophy*. Amsterdam 1997: Rodopi, pp. 145—160; Alexius Meinong and Polish Philosophy [1996], *Axiomathes* **7** (1—2), 241—266; On Tadeusz Czeżowski's Semiotic Views [1998—1999], *Brentano Studien* **8**, 37—51; Trouble with Ontic Categories or Some Remarks on Tadeusz Czeżowski's Philosophical Views [2000]. In: T. Czeżowski, *Knowledge, Science, and Value. Poznań Studies in the Philosophy of the Sciences and the Humanities*, vol. 68, pp. 255—267. Amsterdam: Rodopi.

PART I. BEING AND ESSENCE

1. ON WHAT SEEMS NOT TO BE

1.1

There is probably no philosopher who would pass impassively by the problem of what there is. On the other hand, there is more than one philosopher who takes delight in giving a lot of thought to what there is not.

Marian Przełęcki belongs to the latter group. According to his paper "On What There Is Not" [Przełęcki 1981] there is not what there is not.

As for me, I should consider myself as a member of the former group. It is not easy to decide, however, whether or not my belief that there is somehow that which does not exist, contradicts Przełęcki's viewpoint expressed in the paper mentioned above.

I believe the matter will be cleared up if I directly indicate and bring up for discussion each particular thesis of Przełęcki's paper, which I regard as questionable. Thus my gentle and patient reader will be able to judge for her/himself whether or not there is a consensus between Przełęcki and myself on the matter of the non-being of non-being.

1.2

At first sight, it seems to be obvious that any utterance always has a meaning (i.e. is meaningful) provided that it is not internally contradictory. However, there are utterances that do not have any meaning not because they are internally contradictory but rather because we do not understand them at all. Moreover, one cannot claim that only utterances that are not internally contradictory are meaningful. For we do understand internally contradictory utterances. The identification of having a meaning with the lack of internal contradiction does not suit our linguistic intuitions.

We should instead agree with the view that if any utterance is true or not true (i.e. false), then it has a meaning. But the question arises at once when a utterance is true or not true. Past thinkers answered this question usually in such a way: a utterance is true if it refers to a being; it is not true if it refers to non-being. Now they would rather say: it is true if it refers to an existent state of affairs; it is not true if it refers to a non-existent state of affairs.

If a being and non-being, an existent and not-existent state of affairs are anything at all then it is possible to recognise referring to something as a sufficient condition of meaningfulness.

The utterances that are true or not true are propositions. Przełęcki speaks about them when he says: "The sentence as a whole — not being a name[1] — does not denote, or refer to, anything. Its semantic function consists in something else: in being true or false, not in referring" [Przełęcki 1981, p. 129]. He does not want to treat referring to something as a sufficient condition of meaningfulness. The main reason is what Willard van Orman Quine has called "Plato's beard". If *every* proposition were to refer to something then false ones would also refer to something, and to what if not to something as mysterious as non-being? One must not ask what a false proposition refers to. We ought to be satisfied with an answer to the question *when* a utterance is false. The logician answers this question in the following way: The proposition "*a* has the property *A*" is true *iff a* has the property *A*. Does such an approach allow us to say "about it all that we intend to say" [Przełęcki 1981, p. 126]?

I wish to put the problem of epistemic value of such a statement aside. But two matters are essential here.

First, if one accepts such a condition of truth then the following utterance seems to be correct: The proposition "Swaróg's[2] statue has been profaned" is true *iff* Swaróg's statue has been profaned. It seems to be as correct as the utterance: The proposition "Mściwój[3] has in-

[1] In the Lvov-Warsaw terminology, the term "name" is an equivalent of the English "description" (or "noun phrase").

[2] Swaróg is one of the most important divinities in the Slavonic pantheon.

[3] Mściwój is the name of one of the princes of Polish Pomerania in the late medieval period.

sulted Swaróg" is true *iff* Mściwój has insulted Swaróg. And it is not of the slightest consequence that Swaróg's statues were real chunks of wood or that one could (for example) burn them to ashes, whereas it would be pointless to look for Swaróg among real objects. Such utterances are not immediately entangled in the problem of the existence or non-existence of anything.

Second, if one recognises such a determination of the conditions of truth as satisfactory, one can be satisfied with the following answer to the question of when a utterance has a meaning as well: A given utterance has a meaning *iff* there is another (i.e. of a different form) utterance synonymous with it. It is quite possible that we will never go further than this statement does.

1.3

The word "non-being" can appear in the sentences of the form "*a* has the property *A*" in place of "*a*" as well as of "*A*". Let us consider the following utterances:

(1) Non-being is perceptible.

(2) Swaróg is a non-being.

According to common sense, both utterances have a meaning. Moreover, the former is false whereas the latter is true.

But the logician says: "If *a* is *A* then there is *a*". The sentence (1) is false *iff* the sentence "Non-being is not perceptible" is true. Since non-being is not perceptible then there is non-being. What does the word "non-being" name when it is used as a singular name? In such a case "non-being" appears to name what there is not (the totality of what there is not). The consequence of the statement (1) is the statement that there is what there is not. If someone does not believe that an internally contradictory utterance has no meaning (i.e. is meaningless) then s/he has no reason to think that the utterance "There is non-being", and thus the sentence (1), is meaningless. It would be too bad, however, if some true sentences implied an internally contradictory utterance.

How to escape such a consequence?

Przełęcki mentions three ways of escaping them [Przełęcki 1981, pp. 125-126]: by replacing the name "non-being" with the phrase "what there is not" (Bertrand Russell's way), or the predicate "is not" (Quine's way), or finally by recognising the name "non-being" as an empty name (Jaakko Hintikka's way). Each solution saves us from ascribing being to what there is not but at a cost which Przełęcki would not want to pay.

The logician says: if *a* is *A* then there is *a*. The sentence (2) is true if "non-being" means here the same as "something that there is not". Since Swaróg is a non-being then there is Swaróg. Thus admitting the sentence "Swaróg is something that there is not" we should also admit the sentence "There is Swaróg". The first two solutions (Russell's and Quine's) imply admitting the falsity of the sentence (2), and the third way (Hintikka's) leaned to its acceptance, that Swaróg is a non-existent but possible object. Thereby the ghost of "a highly enigmatic notion of not-existent objects" [Przełęcki 1981, p. 126] appears. According to Przełęcki, one would be free of such a consequence by giving the word "non-being" the meaning "not-being something". Swaróg is a non-being, i.e. he lacks a certain property. But in such a case everything is a non-being. For is there something that one can truly (in conformity with the facts) apply every predicate to? Such a consequence is perhaps in conformity with Plato's view. But let me repeat: is this truly "what we intend to say" when we say "Swaróg is a non-being"?

1.4

By means of the inscription "(*Ex*) (*Px*)" the logician wants to express the thought that at least one object belonging to *a previously fixed object domain* (i.e. the universe of discourse) has the property *P*. By means of the inscription "(*x*) (*Px*)" s/he wants to predicate that very property of each object belonging to this domain. Both expressions *assume* that the range of variables is fixed beforehand. Let us agree that we shall say about the objects mentioned (i.e. indicated by naming) after the phrase "Let it be a domain consisting of" that they *occur*. Let us reserve the word "to exist" only for predicating of objects that are spatio-temporal. It is obvious that although it is true of each object belonging to the domain under our consideration that it *occurs* (in this domain), it need not be true that every object of this kind must exist (i.e. be spatio--temporal).

It may happen that we take into account a domain *D*, which consist only of spatio-temporal objects. Then, but only then, the following will be true: $(x)\ C\ (x \in D)\ (x$ exists). Membership itself in an object domain of any language is not identical with existence (spatio-temporality), though it can be, in some respect, convenient to consider domains homogeneous with respect to the mode of being, domains containing only spatio-temporal objects in particular.

In general, however, we can say only that *everything* in a domain *occurs* in this domain; we cannot say that everything in a domain exists. We can construct the domain of spatio-temporal objects but we can also construct a domain of objects to which somebody pays reverence, irrespectively to whether they have existence or have not.

It is common to read the inscription "$(Ex)\ (Px)$" as "There is an object *x* such that *x* is *P*". It is admissible when the phrase "there is" is used as a synonym of the phrase "occurs in an (established) domain" not as a synonym of the phrase "is spatio-temporal". Thus, the following paraphrase is not admissible: *a* exists *iff* $(Ex)\ (x = a)$. Admitting the expression "$(Ex)\ (Px)$" does not force us to grant existence (i.e. spatio-temporality) to something that is *P*. Accordingly, the phrase "$(Ex)\ N$ (x exists)" is internally contradictory *iff* it expresses the thought "There occurs in an established domain an object *x* such that *x* does not occur in this domain" or the thought "There exists (i.e. is spatio-temporal) an object *x* such that *x* does not exist (i.e. is not spatio-temporal)". If we do not limit the universe of discourse formed above to spatio-temporal objects then we can admit, without the threat of falling into contradiction, that there occurs in an (established) domain an object *x* such that *x* does not exist (i.e. is not spatio-temporal). This will be the case if, for example, Swaróg *occurs* among the objects of such a domain.

Sometimes we use just such a language where objects like Swaróg belong to the domain of this language and where we allow the truth of propositions like "Mściwój has insulted Swaróg". Then, of course, one will be able to say that «the one who has received imprecations» *occurs* (in the object domain of a natural language) as well as that the one who has insulted (without quotation marks because s/he believes in existence of the receiver of imprecations) occurs. However, the user of that language has no obligation to claim that since Mściwój exists Swaróg also exists.

However, when ontologists tackle the riddle of being, they are not concerned with the question of what one is free to speak about or one must not speak about. We may speak about everything. For what could it be that we would not be able to speak about? But not everything which one may speak about exists. What they have in mind when they discuss modes of existence is modes of *being* (existence in our sense is only one of them), not modes of *occurrence*. There is only one «mode» of occurrence: membership in an object domain of a language. I think that linking together the controversy about «philosophical» existence about the modes of being, with the problem of interpreting the logical notation, is a sort of misunderstanding. Hence, when logicians use the term "empty name", it is not always easy to decide what they have in mind. For if we treat names like "divinity" and "werewolf" as empty names, we mean either that:

(a) objects that are divinities (B) or werewolves (W) do not occur in the domain D of a given language:

$$N\ (Ex)\ K\ (x \in D)\ (Bx) \text{ or } N\ (Ex)\ K\ (x \in D)\ (Wx),$$

or that:

(b) divinities and werewolves do not have a certain mode of being, e.g. they do not exist:

$$(Ex)\ K\ (x \in D)\ N\ (Bx) \text{ or } (Ex)\ K\ (x \in D)\ N\ (Wx).$$

However, Swaróg, for example, can belong to the denotation of the name "divinity" — although he does not belong to the denotation of the name "werewolf".

If one comes to agreement — in every case — as to the kind of emptiness under consideration then the Platonian dilemma will disappear: it will become a closed problem.

1.5

The only answer to the question of what exists can be given by indicating the criterion that must be fulfilled by a given object belonging to the domain under consideration, i.e. occurring in this domain, in order to be recognised as an *existent* object. Such a criterion is the possibility of being experienced. Something exists *iff* it may be experienced.

Sometimes it is convenient to assume that each object occurring in any object domain of a given language exists in some way: it implies the necessity of distinguishing real existence from unreal existence. Let us keep the old word "existence" to mean exclusively only real existence (i.e. spatio-temporality). Thus, we are free to claim that every object that occur (in any object domain of a language) *is in a certain mode of being* but not every object is real (i.e. exists); some objects are non-real. The answer to the question of what is (i.e. has being) will be as follows: Something is (has being) *iff* it is determinable (i.e. can be determined at least by means of one predicate).

Moreover, one can stipulate that objects that are determinable but can not be experienced are purely intentional or mental objects. Werewolves and Swaróg mentioned above and at least part of «Plato's beard», in all likelihood would be such mental objects.

Of course, many objects do not exist but one cannot say that an object has no being at all. If one says something about a non-existent object, one thereby gives it mental being.

Among mental objects, philosophers distinguish such objects whose determination (or description) does not contain contradiction and calls them "possible objects". They can also isolate among the latter those objects whose description does not contradict exclusively logical laws, i.e. logically possible objects, as well as those objects whose description does not contradict physical laws, i.e. physically possible objects. I doubt that it is possible to analyse the mental-object domain beyond stating that some objects are possible while others impossible. But this does not show that these objects are philosophically suspicious nor does it demonstrate that the talk about them is logically dangerous. After all, what seems to be essential in philosophical dispute about the problem of being is the degree of the possibility of being of any object «by itself» — and not the mode of being of several objects (or the world in general) or the existence of them in particular.

The controversy among ontologists is the controversy about the *degree of dependence* to which existent objects or the real world as a whole are entitled.

1.6

Summing up, then, the views expounded above are not quite new — especially in the Lvov-Warsaw School. As a matter of fact, all of them have been already formulated by thinkers of this School, more than once better. However, as one of representatives of the Lvov-Warsaw School, Marian Borowski, said about his thoughts in similar circumstances, "it is their literary defect but material virtue" [Borowski 1921, p. 140]. Thus it might be useful if I put my beliefs once more in a nutshell:

1. No logical formula forecloses metaphysical problems.
    1.1. Internally contradictory expressions are meaningful.
    1.2. Sentences refer to states of affairs (situations).
    1.3. The object domain, which any language refers to, is not identical with the real world.
        1.31. Formula "(*Ex*) (*Px*)" should be read as follows: "A certain *x* has the property *P*" or "There is a certain name, which after being substituted into the expression "*Px*" makes it a true sentence".
        1.32. The truth of the sentence like "*a* has the property *A*" — and its meaningfulness — neither assumes, nor implies (real) existence of the object *a*.
            1.321. There are true sentences concerning non-(really)existent objects.
        1.33. Sentences of the form "*a* exists" are meaningful.
    1.4. One can give any metaphysical interpretation of the semantic definition of truth.
2. The correct formulation of metaphysical problems requires introducing the notion of various modes of being.
    2.1. There is a difference between "being" and "existence".
        2.11. "Existence" is a predicate.
        2.12. "To exist" means the same as "to be real".
            2.121. "To be real" means the same as "to be spatio-temporal".
            2.122. The criterion of reality is the possibility of being experienced.

2.123. "To be unreal" means the same as "to be only determinable".

3. Every object is (i.e. has being) but not every object exists.
4. The correct formulation of the metaphysical problems requires distinguishing between two meanings of "non-being".
   4.1. "Non-being" in the first sense means the same as "that which is not something".
   4.2. "Non-being" in the second sense means the same as "nothing".
5. There are only objects (i.e. beings); there is not «non-being» (in the second sense of the word).
   5.1. There is only one empty name: "nothing".
6. The main task of metaphysics is the analysis of the degree of dependence among real objects.

I will be very content if I can make my Readers consult the works of the thinkers of the Lvov-Warsaw School and give some thought to all these problems anew. If my Readers were to take the view that *my* views are incorrect, I would seek my defence in these thinkers' work. "Existence has always been a much discussed problem", as Tadeusz Czeżowski [1964, p. 72], one of the students of Kazimierz Twardowski, wrote. "I hope that the ideas presented above contribute to this discussion."

2. ON THE CONTROVERSY ABOUT UNIVERSALS

2.1

The main problem of the controversy about universals is their ontic status. Solutions to the controversy that have been advanced depend on the ontological positions held by participants in the controversy and on top of this these solutions influence the semiotic standpoints of the thinkers involved.

Participants in the controversy, especially in ancient times, have very seldom made clear how they understand the word "universal". However, one can conclude from their various utterances that they would agree with the following definition:

U is a universal in relation to all particular objects *R* that possess a complex of common properties *C iff U* is an object determined by a complex of properties *C* common to all particular objects *R*.

"A complex of properties *C* common to all particular objects *R*" is meant here as "a complex of properties *C* shared by every particular object *R*". It is ambiguous whether or not *U* is determined solely by the mentioned complex *C*. The phrase "to be determined by a complex of common properties" is understood in various ways, according to an accepted solution to the matter of the ontological status of universals.

2.2

The views of the participants in the controversy vary on the question of whether or not there are objects other than particular objects, i.e. individuals. The latter are considered to be spatio-temporal (i.e. real), separate (i.e. concrete) and autonomic (i.e. objective). So the question is whether or not there are non-spatio-temporal (i.e. ideal objects or, in short, ideas) that are non-separate (i.e. abstract objects or abstracts) and non-autonomic (i.e. subjective objects or subjectives).

The most frequent points of view are those that admit individuals (e.g. Jan Sobieski) as well as ideas (e.g. man in general), abstracts (e.g. Jan Sobieski's sunburn, Jan Sobieski's being older than Kara Mustafa, the class of Turks defeated by Jan Sobieski's troops during the battle of Vienna) and subjectives (e.g. Jan Sobieski's thought of victory) or some of these categories (in various combinations). But some thinkers recognise ideas or abstracts or subjectives while refusing individuals the status of an ontological category.

Particular positions in the controversy are in general justified by an appeal to epistemological criteria. The difference springs from various estimations of human cognitive activity: perception, generalisation (or ideation), and abstraction. Some of the thinkers accept the validity of all these kinds of activity, other thinkers only some of them, and some others reject the validity of all human cognitive activity and recognise it only as a fully creative activity, i.e. on a level with fantasy.

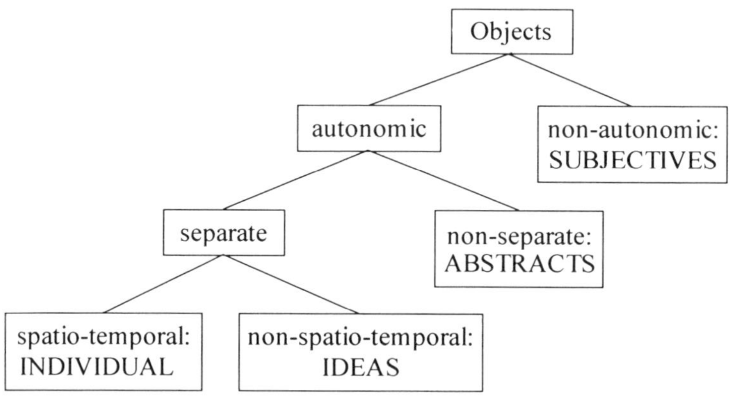

Fig. 1. Categories of objects

2.3

There are four main views concerning the ontic status of universals.

According to the first view — realism — universals are ideas (*universalia ante res, universalia separata*). The phrase "to be determined by a complex of common properties" which is a part of the *definiens* of the general definition of *universal* means here the same as "to be constituted by a complex of common properties". It means that the universal (i.e. idea) in relation to all particular objects (i.e. individuals) *R* possesses the complex of properties *C* common to all particular objects *R*. Thus, for example, the complex of properties common to all men (Jan Sobieski, Kara Mustafa, etc.) is vested in the idea of Man.

According to the second view — conformism — universals are abstracts (*universalia a parte rei, universalia in rebus*), i.e. properties or sets in particular. The phrase "to be determined by a complex of common properties" means here the same as "to be identical with complex of common properties" or "to be identical with all particular objects possessing a complex of common properties". So the universal (i.e. abstract) in relation to all individuals *R* either is identical with the complex of properties *C* common to all individuals *R* or is identical with all of them, i.e., with a set of all individuals *R*. Then the abstract of Man is the complex of properties common to all particular men or simply a set of all particular men.

According to the third standpoint — conceptualism — universals are subjectives (*universalia post res*), i.e. conceptions or notions. The phrase "to be determined by a complex of common properties" means here the same as "to have attributed a complex of common properties" or "to be identical with an attributed complex of common properties". So the universal (i.e. subjective) in relation to all individuals *R* either possesses the attributed complex of properties *C* common to all the individuals *R* or is identical with this attributed complex of properties *C*. The subjective of Man either is, for instance, the object to which the complex of properties common all particular men is attributed — i.e. it is the conception of man in general — or it is identical with this attributed complex — i.e. it is the notion of man in general.

Finally, according to the fourth view — nominalism — universals are individuals (*res*), i.e. names. The phrase "to be determined by a complex of common properties" means here the same as "to connote a complex of common properties". Thus, the universal (i.e. individual) in relation to all individuals *R* — thanks to its connotation — denotes all individuals *R*. The individual (i.e. name) of Man denotes all particular men.

2.4

Admitting a certain ontic status to universals implies a definite solution to the problem of semantic categories of expressions, semantic categories of names in particular. In the controversy about universals, three semantic functions of names are mentioned. Names can designate a referent, i.e. a *designatum*, connote a content, i.e. a *conotatum*, and denote an extension, i.e. a *denotatum*.

For a long time there have been differences of opinion concerning the questions of whether names perform all the above-mentioned semantic functions and whether these functions are performed by all names. Some philosophers grant that names — or universal names at least — perform all indicated semantic functions. They contend that the referents of universal names are individuals, and the referents of abstract names are abstracts. Then they identify the extension with an abstract (i.e. a set), and the content either with an abstract (i.e. a complex of properties) or a subjective (a conception or a notion).

Some semioticians think that we can fully describe the role that names play without attributing to them the function of connoting a content. They take it for granted that names (universal names at least) perform only the function of designating and denoting. They contend that the referents of singular names are individuals and those of abstract names are abstracts. According to this opinion, extensions of universal names are either individuals (i.e. collectives) or abstracts (i.e. distributives). Thus, all terms concerning a content (i.e. a sense) of the name can be replaced by expressions without the term *content*. For instance, one can say that a certain name N_1 is synonymous with a certain name N_2 instead of saying that a certain name N_1 connotes a certain defined content.

According to another view, names perform one semiotic function only, namely the function of predicating, identified with designating. Then again, some semioticians divide names into two semantic categories: referents of singular names have to be individuals and those of universal names have to be universals — *viz.* ideas or subjectives (i.e. conceptions). Other semioticians take it for granted that there is just one category of names. They claim that the referents of all names are individuals but names can be used to designate many individuals one after another.

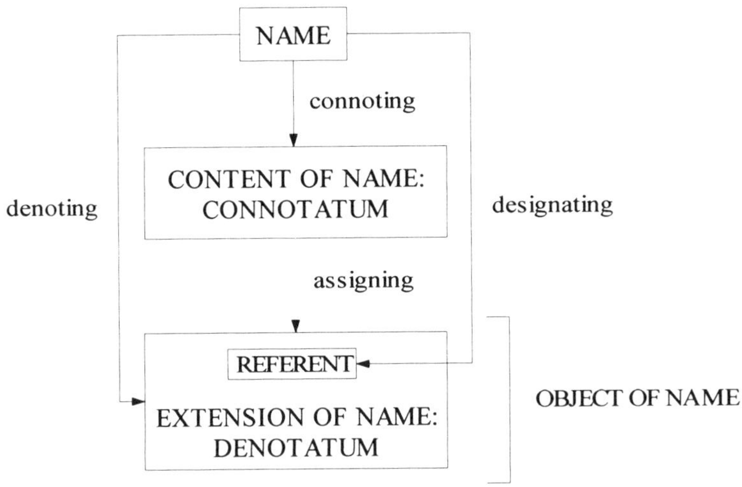

Fig. 2. Semantic functions of names

2.5

The controversy has been so enduring due to the fact that many participants have not: (1) tried to construct a logically correct definition of a *universal*; (2) been aware of the ontological premises that underlie particular views; (3) known how to separate semiotic problems involved in the controversy from ontological ones; and (4) made distinctions among particular semantic functions of names. For these reasons, the correct formulation of the views pronounced by participants in the controversy requires an operation of explication, with all its limitations and consequences.

3. ON FORMS OF OBJECTS

3.1

Many years ago, in Cracow, Roman Ingarden wrote in his paper "Was wir über die Werte nicht wissen": "*Ich will mir gewiße Schwierigkeiten, auf welche die Werttheorie stößt, zum Bewußtsein bringen, um Wege und Weisen ihrer Überwindung zu suchen*" [Ingarden 1964, p. 97]. The intentions of this paper are very similar. I am far from a "*skeptische Hochmut*" in approaching the problem of forms of objects characteristic of the work of some scholars who recognise it as a scholastic pseudo-problem bereft of topical interest. But I must also admit that I am distant from a minimalistic humility, which would lead me to consider this problem as one of the irresolvable mysteries of the world.

3.2

To begin with: I assume:

(1) For every x [x is an object *iff* for a certain P (Px)].

Formula (1) is increasingly gaining acceptance [Kotarbiński 1931, p. 70; Augustynek 1984, p. 3]. It is remarkable that reists, who also assume this formula, deny at the same time the thesis that, for example, properties belong to objects [Kotarbiński 1930—1931, p. 200; 1931, p. 73]. As Kazimierz Ajdukiewicz writes, the opponents of reism would appeal to the fact that it is not only things that we can predicate properties of, whereas reists would deny this fact.

Can such a controversy "be settled within ordinary language" [Ajdukiewicz 1934d, p. 107]? In fact, reists can only make use of the semantic postulate that an object is a body [Kotarbiński 1929, p. 55] and so (according to another postulate), it is something inert.

Let us consider the question more closely. Let us contrast, in particular, the following words from natural language: "thing", "event", "change", "state", "property", "relation", "set", "part". These words do not have precise senses,[1] but I shall try to compare them within the bounds of their sharpness.

First, I would like to point out that some of these words can be iterated; natural languages allow us to talk about properties of properties, properties of properties of properties ...; about relations between relations ...; about sets of sets ...; about parts of parts, and so on. It is inadmissible, however, to talk about things of things, or about events of events. The question of the admissibility of talking about changes of changes ..., and about states of states ... does not have a clear answer. Let us agree, that:

(2) For every x and y (x is a property of y *iff* y is the subject of x).

(3) For every x, y and z (x is a relation between y and z *iff* y and z are members of x).

(4) For every x and y (x is a set containing y *iff* y is an element of x).

(5) For every x and y (x is a part of y *iff* y is a complex including x).

My first remark is that properties can be subjects of properties, relations can be members of relations, sets can be elements of sets, and parts can be complexes of parts. Secondly, iterable terms can be linked together and they can be added to the other terms. Let us call this phenomenon "interpredicativity". It is admissible to talk about properties of relations, sets, and parts, as well as about properties of changes and states, and about properties of things and events; we can talk about relations between properties, between relations, between sets, and between parts, as well as about relations between things and between events; it is admissible to talk about sets of properties, relations, and parts, as well as about sets of things, and events, and changes, and states. We can probably also talk about parts of properties, about parts of sets, about parts of things, and about parts of events, as well as parts of changes and states, but hardly about parts of relations.

---

[1] Even serious philosophical texts inexactly talk about "properties, relations, dispositions, events etc.", or about "properties, relations, states, processes etc." (cf. for instance [Kotarbiński 1931, p. 74] and [Ajdukiewicz 1948, p. 69]).

The terms "change" and "state" can also be added to at least some of the remaining terms. Thus, it is admissible to talk about changes in properties, relations, parts, and things, but hardly about changes in sets and events; and it is more difficult to talk about states of properties and relations. I doubt whether it is admissible to talk about states of sets, but we certainly can talk about states of parts.[2] Using the convention introduced above and accepting the view that things, events, changes, states, properties, relations, sets and parts are, in any case, objects, my second remark is that all objects can be subjects of properties, members of relations and elements of sets but only some objects can be complexes of parts. Similarly, we cannot say about any object that it undergoes changes or that it is *in* a certain state (of affairs).

3.3

Are there any ontic relations that correspond to the above syntactic relations between terms? If so, we are forced to accept the view that two separate forms of objects can be distinguished: the form of things and events, on the one hand, and the form of properties, relations and sets, on the other hand. Following Aristotle (cf. [Kotarbiński 1929, p. 40]), let us call objects of the second form "accidents" (or "fortuities").

We have now:

(6) For every x [x is substantial *iff* it is not the case that for a certain y (x is an accident of x)].

(7) For every x [x is accidental *iff* for a certain y (x is an accident of y)].

The possibilities of iteration and interpredicativity are not the only pecularities of accidents. All accidents are also (contextually) reducible one to another. Let us suppose that A stands in relation S to B. If so, then A has a certain property: the property of standing in relation S to B. And if so, then A belongs to a certain set: the set of objects that have the property of standing in relation S to B. Since A belongs to this set, A has thereby also a certain property, *viz.* the property of belonging to this set. Furthermore, since A has the property of belonging to a cer-

[2] Augustynek discusses this probability in this context [Augustynek 1975, p. 131].

tain set, *A* stands in a certain relation to *B*, *viz.* the equality-relation with respect to belonging to the set to which *B* also belongs. This reducibility also finds its expression (on the logico-philosophical grounds) in the identification of properties and relations with respective sets. The question thus arises: Is also the operation of distinguishing properties, relations and sets among accidents a purely verbal operation?

Let us now consider the status of changes and states (of affairs). "Change of *x*" means, in the first sense, the same as "replacing (at a certain time) one property of *x* with another one"; in the second sense it means the same as simply "replacing". We use the word "change" in this last sense, when we say, for example, that a movement (of any body) is a change of locality (of this body). Likewise, "state of *x*" denotes, in the first sense (let us call it "the situation-sense"), a lasting (at a certain time) of a property of *x*; in the second sense (let us call it "the attributive sense"), it denotes a lasting (in a certain sense) property of *x* — i.e. something which, as Henryk Stonert says, a given thing is in [Stonert 1964b, p. 169]. Tadeusz Kotarbiński was probably aware of this polysemy when he identified states of affairs and changes (processes), first, with the fact *that* things happen in such-and-such way or with the fact *that* they change in such-and-such a way [Kotarbiński 1954, p. 395] and, later, with *how* things happen and *how* they change [Kotarbiński 1935, p. 118—119]. Bogusław Wolniewicz points out a certain ambiguity in the Polish equivalent of the English expression "state (of affairs)" which has its roots in the following amphibology: the idea here can be not only of a state of *x* (or a state in which *x* is present), but also of a state of *x*-s (or a mutual configuration of many things) [Wolniewicz 1968, p. 95], that is, of the existence of a certain relation between *x*-s. He proposes paraphrasing the second sense with the aid of the phrase "that it is so-and-so" [Wolniewicz 1968, p. 319].

Every change is a certain «situation»-state [Stonert 1964a, p. 169]. While a «situation»-state is a certain event, an «attributive» state is a certain property. Not every property, of course, is an (attributive) state; only a relatively unstable property is involved. Thus, extra-temporal properties are not in question here (if such properties exist) do not constitute states. Ingarden also makes a distinction here, though he does so in a different way: "*Zustand muß mehr oder weniger dauernd sein*" but it must be something (in Ingarden: "*der Gesamtbestand*") that has been created "*in dem Gegenstande durch einen Vorgang*" [Ingarden 1947, p. 224]. Similarly, not every event is a change; only events that are capable of happening exclusively in a certain place come into play here; thus, extra-spatial events are not in question here (if such events exist). Let us suppose that a buried in thought thin man is walking. We shall say, that thinness is his *property* but that he is in the *state* of contemplation, and that he undergoes a certain *change* (here: of location). When something rots, we can say that it undergoes a certain *change* (i.e. rotting), that it is in a certain *state* (i.e. of rotting), and that it has a certain *property* (i.e. of rotting). When somebody is irritated, we can say that s/he is in a certain *state* (i.e. of irritation), that s/he has a certain *property* (i.e. of irritation or, better, of being irritated), but we cannot say that s/he undergoes a *change*. When something is green, we can say that it has a certain *property* (i.e. of greenness) but we cannot say that it undergoes a certain *change*, or that it is in a certain *state*. Moreover, when something rots, when somebody is irritated, or when something is green, we can say that this object belongs to a certain *set* (i.e. of rotting, irritated, or green objects, respectively). But when a certain object is a man, we do not usually say that this object *as such* undergoes a certain *change*, or that it (*as such*) is in a certain *state*, or that it (*as such*) has a certain *property*. Thus, perhaps it would be possible to reduce the category of accidental objects (i.e. accidents) to objects of just one kind, to sets, say.

3.4

I have included things and events among substances. I might leave it at that (as does Peter F. Strawson [1959, p. 15]).[3] But we can inquire as to whether the reduction outlined for accidents can be performed for substantial objects as well.

We may therefore ask whether things can be reduced to events, or *vice versa*. It seems that the pair part-complex could be a mediator in both cases. Then, correspondingly, the term "thing" would mean the same as "complex of events" ("*Verband von Sachverhalten*") [Ingarden 1948.1, p. 284] or "part of an event"; and the term "event" would mean the same as "complex of things" or "part of thing". Assuming

[3] In fact, Strawson considers things and events as particulars.

that complexes of events and parts of events are events by themselves, and also that complexes of things and parts of things are things, we may talk about two ontic forms only: events (or things) and sets (or properties, or relations).

The situation would be different if things were reduced to events (or *vice versa*) by means of the element-set pair. Tadeusz Czeżowski mentions such a procedure. Things ("individuals enduring in time") are thus sets of events ("momentary individuals"), and the latter, as "objects of a higher logical type" [Czeżowski 1951, p. 220], are not events.

To assess the admissibility of such a "clean-shaven picture of reality", as Bertrand Russell described it [Russell 1959, p. 66], we should make a prior choice as to the specific explication of the terms "thing" and "event" since they are far from being precise in natural languages. It is impossible to do without arbitrary decision here. Some authors answer the question of what things and events are by indicating the requisite objects — not so much by ostension [Augustynek 1984, p. 4] as through exemplification by means of verbal tools — the required objects. However, these exemplifications, as well as the comments appended to them, are usually rather imprecise.

3.5

Let us consider THINGS, first. "Things include: tables, stones, trees, houses, men [...]." Socrates is certainly a thing, but what kind of object is Socrates? Andrzej Grzegorczyk proposes the following quasi-grammatical criterion: things are identical with "designates of the majority of nouns" [Grzegorczyk 1959, p. 10]. But this criterion is of little use here; after all, grammarians usually answer the question "What is a noun?" by saying that it is a name of things, primarily.

Let us assume for the sake of simplicity that, prior to Socrates' birth, Socrates was nothing at all and that, after his death, he is fully annihilated and becomes nothing at all (this assumption is, of course, far from correct). In every period of his life, Socrates has various properties, including various states and changes. Let us ignore the fact, again for the sake of simplicity, that he is also (willy-nilly) the member of various relations. Is Socrates (a) the «residuum» that remains after separating these properties, or (b) this «residuum» with all these properties, or (c) this «residuum» only with those («essential») properties that are not states? The following counterintuitive consequence speaks against version (a): all sentences that stated something true about Socrates would be synthetic (moreover, would such a «bare» Socrates be an object at all?). Version (b) has the equally counterintuitive consequence that all true sentences about Socrates would be analytical.

EVENTS are generally identified with "a given occurrence of a phenomenon [here: of a property *in specie*] in a given object and at a given time" [Ajdukiewicz 1965b, p. 163], or with "the fact that an individual located in a determinate place and time has [...] [a certain] property" [Ajdukiewicz 1934b, p. 90] or, more generally, with the fact that a certain thing has a property (or the fact that some relations occur between some things) [Stonert 1964b, p. 53]. This is how the situation is described by Ajdukiewicz and Stonert. It should be remembered that states of affairs (of the «situation» kind) and changes are only some types of such events. According to Kotarbiński, for instance, an event is either a state (static events) or a change (kinetic event) [Kotarbiński 1929, p. 51]; though sometimes he equates an event with a state [Kotarbiński 1954, p. 396]. In Augustynek, we read of the possibility of identifying events and changes (processes) [Augustynek 1975, p. 95]. According to Russell who uses here the term "fact" after all events "consist always of relations between parts of a whole or qualities of single things. [...] It is convenient to use the word "fact" to express the analysed connection of the parts rather than the complex whole that they compose" [Russell 1959, p. 151]. Some authors give grammatical criteria for being an event or being a fact as well. George Edward Moore, for example, writes: "I am going [...] to use the name "facts" simply and solely as a name for [...] the kind of things which we express by phrases beginning with "that"" [Moore 1953, p. 298]. Wolniewicz distinguishes events (states of affairs) from facts. The latter are "existing states of affairs" [Wolniewicz 1968, pp. 122-123] or rather "the existence of states of affairs", i.e. they are what is stated by a true sentence [Wolniewicz 1968, p. 97]. On the other hand, according to Ingarden, events are identical with "*das Ins-Sein-Treten eines Sachverhalts*" [Ingarden 1947, p. 216], where these states of affairs can be «*handlungsmäßig*» or «*eigenschaftlich*» [Ingarden 1948.1, p. 315].

In these circumstances, many scholars construct a technical sense of the word "event", without looking at its use in natural languages. For Ryszard Wójcicki, for example, events are "momentary sections of a thing" [Wójcicki 1972, p. 21]. For Hans Reichenbach, events "are space-time coincidences and do not endure" [Reichenbach 1947, p. 267]. In Karl Popper, an event is not "a complex, or perhaps a protracted, occurrence, whatever ordinary usage may suggest" but a class of occurrences. Thus, for example, that a glass of water has just been upset (i.e. a certain occurrence) belongs to the event that consists in "the upsetting of a glass of water" as to a certain event [Popper 1934, p. 89].

I would add that processes (changes) are sometimes reduced to events [Augustynek 1975, p. 73] or (momentary) states of affairs [Quine 1950, p. 67] as sets (or complexes) of these events or states of affairs, partly ordered by the relation of anteriority.

3.6

Looking for objects, to which the remaining ontic forms can be reduced is not the same as looking for basic (atomic) objects.

I assume:

(8) For every *x* [*x* is atomic *iff* it is not the case that for a certain *y* (*y* = *x* and *y* is a part of *x*)].

(9) For every *x* [*x* is molecular *iff* for a certain *y* (*y* = *x* and *y* is a part of *x*)].

The division of objects into atomic and molecular is probably feasible within the realm of substances, and probably also within the realm of accidents. Russell identifies molecular objects with facts. He writes: "facts [...] are whatever there is except what (if anything) is completely simple" [Russell 1959, p..151]. Rather unexpectedly, Ludwig Wittgenstein calls objects that fulfil conventions (8) "things". Wolniewicz identifies Wittegnsteinian things with material points and notes that the Wittgensteinian criterion "does not imply [...] individuality (concreteness) of his objects" [Wolniewicz 1968, p. 78].

It is worth stressing that if the search for individual (atomic) objects did not involve partitioning (i.e. isolating parts) but abstracting (i.e.

isolating accidents), this would impose the condition of not having any accidents (properties in particular) on atomic objects. This would give rise to the paradoxical situation, in which atomic objects would not be objects at all, according to formula (1).

3.7

I shall further distinguish the division of objects into substantial and accidental, on the one hand, from the division of objects into autonomous and heteronomous, on the other.

I assume:

(10) For every x {x is autonomous *iff* it is not the case that [if for a certain y ($y = x$) then for a certain z ($z \neq x$)]}.

(11) For every x {x is heteronomous *iff* [if for a certain y ($y = x$) then for a certain z ($z \neq x$)]}.

These formulae are explications of the phrases: "primary existence", "independent existence", "existence in abstraction from anything", on the one hand, and "existence in something" [Ajdukiewicz 1949, pp. 78, 84], "existential dependence" [Augustynek 1975, p. 133], on the other.

If things are complexes of events then things are heteronomous objects. Are events autonomous? Reichenbach claims cautiously that if things turn out to be "classes of events" then "for physics, events are more fundamental units than things" [Reichenbach 1947, p. 267]. Since Russell CONCEIVES "each event as occupying a finite amount of space-time and as overlapping with innumerable other events which occupy partially, but not wholly, the same region of space-time" [Russell 1959, p. 20] then no event is autonomous according to this ASSUMPTION.

Similarly, if events are complexes of things then events, but not things are heteronomous objects. However, we should not jump to the conclusion that consequently only things are autonomous. On the contrary, everything seems to show that they too are heteronomous.

3.8

I now assume:

(12) For every x and y [x is separated from y *iff* for a certain z ($z \neq$ x, $z \neq y$ and z is between x and y)].

(13) For every x and y {if x or y is subjected to suitable forces then [x is separable from y *iff* x (not having been separated from y before the action of these forces) becomes separated from y]}.

(14) For every x [x is concrete *iff* for a certain y (x is separable from y)].

(15) For every x [x is abstract *iff* for every y (it is not the case that x is separable from y)].

Of course, one needs to establish whether this separation of objects consists in a spatial or a temporal separation, or in both (where one must also establish what is to be a level of magnitude), or in something else (the term "separation" should be, in any case, understood in such a way that abstracting, i.e. mentally isolating properties in any object, is not denoted by the term). In the first three cases, abstract objects would be heteronomous as well while concreteness would exclude neither autonomy nor heteronomy.

3.9

The opposition concretes-abstracts is sometimes characterised not by means of separability but by means of separation. To prevent possible misunderstandings, I shall talk in these cases of isolated and connective objects.

Thus:

(16) For every x [x is isolated *iff* for every y (if $y \neq x$ then x is separated from y)].

(17) For every x [x is connective *iff* for a certain y ($y \neq x$ and it is not the case that x is separated from y)].

Of course, all isolated objects are concretes, but some concretes can be connective. Likewise, all abstracts are connective, but some of connective objects can be concretes.

Willard van Orman Quine argues against identifying concretes with objects that are "continuous in geometrical shape"; he points to the fact that, for example, "the territory of the United States including Alaska is discontinuous, but it is none the less a single concrete object; and so is a bedroom suite, or a scattered deck of cards" [Quine 1950, p. 69]. Reichenbach goes further by saying that, according to the spirit of natural languages, "the furniture of a certain house is not an individual, but a class of individuals". Earlier he defines "individual" as "something occupying a continuous and limited part of space and time", and therefore as something that we would determine as just an "isolated object" [Reichenbach 1947, p. 266]. Material concretes which constitute an «equipment» of the world, are characterised according to formula (16) by Małgorzata Czarnocka. She gives a denotative definition of "concretes"[4] and indicates bodies as well as (among others) waves and fields, since she views bodies as being "concentrated in a part of space, close (in general) nubbles of matter with positive mass and of any form" [Czarnocka 1986, p. 14]. This would imply that the world, at least in spatial terms, has a «grainy» structure: it is these «grains» (of various «plies») which fill phase-complexes separated from one another by portions of space(-time)[5] that would be concretes. As Russell succinctly put it, the universe would be "all spots and jumps" [Russell 1931, p. 98].

3.10

I shall now distinguish the pair: concreteness-abstractness from the pair: intelligibility-incompatibility.

I assume:

---

[4] In fact, Czarnocka talks about physical individuals (as well as about empirical or real individuals), but we can interpret her statements as concerning concretes alone, since she contrasts these physical individuals with, i.a., properties, processes, and events.

[5] We should probably agree with Kotarbiński that "empty fragments of space" cannot be concretes [Kotarbiński 1935, p. 488].

(18) For every x [x is intelligible *iff* it is not the case that for a certain P (Px and it is not the case that Px)].

(19) For every x {x is incompatible *iff* [for a certain P (Px and it is not the case that Px)]}.

If we impose no further restrictions on intelligible and incompatible objects, we cannot exclude beforehand that divisions of objects into intelligible and incompatible ones, on the one hand, and into concrete and abstract ones, on the other hand, are logically independent.

3.11

The division of objects into concrete and abstract is frequently connected — or confused — with the divisions of objects into individual and universal, on the one hand, and particular and general, on the other hand. Kotarbiński claims, for instance, that universals are a kind of abstracts [Kotarbiński 1954, p. 359], whereas Augustynek CONSCIOUSLY IDENTIFIES abstracts with universals [Augustynek 1975, p. 64] and then identifies the latter with sets [Augustynek 1984]. Similarly, Quine identifies abstracts with universals, but in regarding sets to be universal objects, he does not identify the former with the latter [Quine 1937—1950, pp. 114—115, 128].

I assume:

(20) For every x {x is individual *iff* for a certain P [Px and for every y (if Py then $x = y$)]}.

(21) For every x {x is universal *iff* for every P [if Px then for a certain y (Py and $x \neq y$)]}.

The difference between individuals and universals is clearly indicated by Joachim Metallmann:

> If we describe two [«identical»] buttons successively in two long series of sentences [...] of such a kind that every sentence distinguishes, say, one property of the object, then among these series of sentences besides identical sentences there will always show up sentences *different* as to their content, provided that the description is sufficiently long. [...] Only [universal] objects can be identical, i.e. such that whatever can be told about one of them, can also be predicated about any other one [Metallmann 1939, pp. 24, 27].

The distinction expressed by formulae (20) and (21) can be carried out for objects characterised in formula (15) as abstracts: these (properties, in particular) happen to be either individual or universal. When Ajdukiewicz calls "a property which at a certain time [can] be an attribute of this or that object, but can also not be its attribute at another time" a "phenomenon", and when he contrasts these "general entities" with realised properties [Ajdukiewicz 1965b, p. 163], he is drawing the distinction between individual and universal abstracts mentioned above. We should bear in mind, however, that "by a "phenomenon" we usually mean the same as by an "event"" [Kotarbiński 1929, p. 79], or — in traditional terminology — "something that is observable" (see below). In any case, as Reichenbach points out, "sometimes events are important units also for the purposes of daily life" [Reichenbach 1947, p. 267].

3.12

I assume:

(22) For every *x* [*x* is particular *iff* for every *P* (*Px* or it is not the case that *Px*)].

(23) For every *x* [*x* is general *iff* it is not the case that for every *P* (*Px* or it is not the case that *Px*)].

Formulae (22) and (23) are explications of the distinction between objects "*in keiner für sie möglichen Hinsicht unbestimmt*" and objects either having "*in dem Gehalte* [...] *eigenartige Elemente*" or being "*nach verschiedenen Richtungen hin ganz unbestimmt*", that is, possessing "*Unbestimmtheitsstellen*" Ingarden calls the former "individual objects" and the latter "ideal" [Ingarden 1947, pp. 40—41] or "(purely) intentional objects" [Ingarden 1948.1, p. 219]. Formula (22) is not, of course, identical with Kotarbiński's assumption that "every object has either a certain property or its negation" [Kotarbiński 1949, p. 422]. Thus, formula (22) describes only particular objects. Metallmann writes:

> The description of such an object by means of individual sentences, enumerating its properties one by one, is strictly speaking never exhaustive. [...] Thus, in everyday life, as well as in scientific practice, the description of an object is always executable, because [...] we break it off in the moment that appears advisable for

> us. But during the evolution of science [...] the number of possible aspects of the «same» object appears to be in principle unlimited [Metallmann 1939, p. 24].

Metallmann calls such an object "concrete" and not "particular". In opposition to the description of *particularia*, the description of *generalia* "can be performed by means of the finite and, in general, small number of properties and relations" [Metallmann 1939, p. 27]. According to Wolniewicz, it is this opposition that is hidden beneath the terms "concrete"—"abstract" in the Hegelian tradition [Wolniewicz 1968, p. 202].

It is rather surprising that things (understood as objects that have only these properties that are not states excluding relations into which these objects enter) are general (as well as heteronomous and, probably, abstract) objects.

Note that under assumption (23) we have:

(24) It is not the case that for every *P* {*Px* or it is not the case that *Px* *iff* for a certain *P* [it is not the case that (*Px* or it is not the case that *Px*) *iff* for a certain *P* (it is not the case that *Px* and it is not the case that it is not the case that *Px*)]}.

Now, there are two possibilities. If we adopt the principle of non-contradiction (in a certain form):

(25) It is not the case that it is not the case that *p iff p*.

then we obtain:

(26) For a certain *P* (it is not the case that *Px* and it is not the case that it is not the case that *Px*) *iff* for a certain *P* (*Px* and it is not the case that *Px*).

Thus, generality is identified with incompatibility. This is the standpoint of, among others, Kotarbiński (who has previously identified *generalia* with *universalia*) [Kotarbiński 1920, p. 14]. On the other hand, if the principle of non-contradiction in the form of (25) is not accepted then the difference between generality and incompatibility will consist in the fact that the former will be ascribed to objects to whom a certain property neither belongs nor does not belong, whereas the latter is ascribed to objects to whom this property both belongs and does not belong. Ingarden thinks that "*das Prinzip der ausgeschlossenen Dritten und das Prinzip des Widerspruchs, in ontologischer Deutung ihre Geltung bezueg-*

lich der IDEENGEHALTE *verlieren*" [Ingarden 1948.1, p. 242], and thus, these principles are characteristic of ideal objects.

Independently of any interpretation, *generalia* understood in terms of formula (23) cannot be identified with sets (in the sense of set-theory). Moreover, it seems that no set is a general object.

Of course, formula (22) does not exclude the possibility that some *particularia* are intelligible whereas others are incompatible.

3.13

In general, two other distinctions of objects interfere with the distinctions discussed above: the division into material and ideal objects, and the division into real and irreal objects.

I assume:

(27) For every *x* [*x* is material *iff* for a certain *y* (*y* is a place-moment, and *x* is at *y*)].

(28) For every *x* [*x* is ideal *iff* it is not the case that for a certain *y* (*y* is a place-moment, and *x* is at *y*)].

In formulae (27) and (28), a "place-moment" means the same as a "spatio-temporal atom".

Kotarbiński initially grants that "material" means the same as "extended AND inert at the same time" [Kotarbiński 1931, p. 71]. However, he provides neither examples of objects that are extended but not inert, nor examples of objects that are inert but not extended. He then identifies materiality with temporality, spatiality (for example, extension in breadth, length and depth) and resistance [Kotarbiński 1935, p. 118]. In the end, according to him "everything that is temporal and spatial and physically defined — for instance, physically influencing something else", is to be material [Kotarbiński 1949, p. 426; cf. also Augustynek 1984, p. 5]. On the other hand, George Edward Moore understands by "material object" something that "is situated somewhere or other in space" [Moore 1953, p. 128], thus ignoring its temporal localisation and also whatever is not "a mind, nor an act of consciousness" [Moore 1953, p. 131].

How are concreteness and materiality related to each other? If the introductory condition of formula (13) concerns a certain material operation then concretes as well as abstracts will be material objects. We can hardly talk about the separation of ideal objects one from another since they are *ex definitione* outside space-time. But if (particular) concretes are expected to be separated one from another, and (particular) abstracts are expected to be unseparated one from another by any portion of space-time, then we cannot exclude that some material objects will be concrete, some will be abstract, and that all ideal objects will be abstract. As Quine notes, "with a little stretch of imagination, [...] [the] notion of physical [here: material] objects can [...] be made to accommodate physical processes or events, on a par with bodies" [Quine 1966—1974, p. 260]. According to Wolniewicz, every event is a material object; in any case, it is "always localised in time". "On the other hand, the content of the word "event" does not imply [...] definitely EXISTENCE" [Wolniewicz 1968, pp. 58ff] (here: objectivity; see below).

Formulae (18) and (27) show that we should not follow the widespread practice [cf. for example, Augustynek 1975, p. 135] of definitionally identifying intelligibility and materiality. On the other hand, these formulae do not exclude the possibility that *de facto* the class of real objects is identical with the class of material objects. The definitional reduction of reality to materiality (or *vice versa*) should, of course, be distinguished from the identification of reality (or physicality) with the possession of spatio-temporal localisation by means of an appropriate meaning-postulate, as Rudolf Carnap [1935, p. 21], Ajdukiewicz [1949, p. 79] and Strawson [1959, pp. 29—30] do (for reality), as well as Quine [1966—1974, p. 259] does (for physicality).

The question arises as to whether the so-called *Universe* fulfils the condition indicated by formula (27). Augustynek claims that it does not fulfil it saying that "in relation to the world as a whole, there is no sense to talk about a spatio-temporal localisation; what is the temporally and spatially external object, which would be the fiducial point of localisation of the world" [Augustynek 1984, p. 5]? Thus, if the possibility of referring to something external is the "immanent property of localisation" then formula (27) should read:

(29) For every x {x is material *iff* for a certain y [y is a place-moment, and x is at y — or for a certain z (z is a part of x, and z is at y)]}.

3.14

I assume:

(30) For every x [x is real *iff* for a certain y ($y \neq x$ and x acts upon y)].

(31) For every x [x is irreal *iff* for every y (if $y \neq x$ then it is not the case that x acts upon y)].

Augustynek appeals to (physical) action in his characterisation of material objects. His definition runs as follows:

(32) For every x {x is material *iff* for a certain y [$y \neq x$ and x acts upon y — or for a certain y and a certain z ($y \neq x$, $z \neq x$; y, and z are parts of x, and y acts upon z)]} [Augustynek 1984, p. 5].

In Ingarden, the condition of reality is causal action [Ingarden 1947, pp. 109-110]. I place this condition on determination:

(33) For every x [x is determined *iff* for every y ($y \neq x$ and x is causally conditioned by y)].

(34) For every x [x is acausal *iff* for every y (if $y \neq x$ then it is not the case that x is causally conditioned by y)].

Regardless of whether we define "reality" or "determination", it seems that the following difficulty arises. If we agree that the range of the variable y is the set of real (*resp.* determined) objects, then we must also use the above formula to decide whether, in a given case, we are dealing with any object at all. Thus, using this formula implies a *regressus ad infinitum*.

In any case, the division of objects into concrete and abstract, on the one hand, and into real and irreal, on the other, are mutually independent on such an approach.

3.15

The pair "concretes-abstracts" is sometimes compared not only with ontological terms but also with epistemological ones — primarily with the pair "observable-noumenal".

I assume:

(35) For every x {x is observable *iff* for a certain y [y is a place-moment and for every z (if z is cognitively directed towards y then z recognises x)]}.

(36) For every x [x is noumenal *iff* for every y and z (if y is a place-moment and z is cognitively directed toward y, then it is not the case that z recognises x)].

Nothing, of course, can be forejudged about the relation of observability and concreteness before we specify what type of cognitive activity is mentioned in formulae (35) and (36). For this purpose, we can assume, for example:

(37) For every x and z (z recognises x *iff* x acts on the receptors of z and z realises that x acts on the receptors of z).

It would be necessary to impose on z the condition of normality (not yielding to delusions) and, probably, reliability (veracity). On what basis, however, are we to admit these and reject other epistemic operations?

Let us suppose that this problem is decided in such a way that observation is limited to extraspection and introspection. If we now assume that we are entitled to talk responsibly only about objects that are observable (in this sense), then for the time being we could not admit the view that some material objects are at the same time incompatible. Nor could we responsibly decide, respectively, the concreteness and intelligibility of ideal object which are simultaneously noumenal.

3.16

I assume:

(38) For every x and z [x is observable for z *iff* for a certain y (y is a place-moment, and if z is cognitively directed towards y, then z recognises x)].

We may now characterise the pair intercognitivity-monocognitivity (or intersubjectivity-monosubjectivity) [Ingarden 1948.2, p. 4].

(39) For every x [x is intercognitive *iff* for every y (x is observable for y)].

(40) For every x [x is monocognitive *iff* for every y and z (if x is observable for y, and x is observable for z, then $y = z$)].

Formula (40) is usually enriched by the condition that x should stand in a determined relation to y ($= z$). Ajdukiewicz expresses this condition by saying, for instance, that "mental phenomena can be perceived only by one person, namely the person who is experiencing these phenomena". He calls this property "intrasubjectivity" [Ajdukiewicz 1938, pp. 157ff]. Although, only mental objects are monosubjective, one should distinguish their monosubjectivity (here: monocognitivity) from their mentality.

I assume:

(41) For every x {x is extra-mental *iff* it is not the case that [if for a certain y ($x = y$) then for a certain z: z consciously experiences x]}.

(42) For every x [x is mental *iff* if for a certain y ($x = y$) then for a certain z: z consciously experiences x].

Ajdukiewicz recognises conscious experiencing (the contents of consciousness) as only ONE of the characteristic properties of mental objects [Ajdukiewicz 1938, p. 158]. His understanding of "objectivity" (here: "extra-mentality") can also be interpreted as follows:

(43) For every x [x is extra-mental *iff* it is not the case that for a certain y (y consciously experiences x)].

Thus, Ajdukiewicz claims that phenomena are objective (here: extra-mental) *iff* "they are never the contents of consciousness and they can exist independently from the consciousness of anybody" [Ajdukiewicz 1938, p. 158].

If we agree, that only experiences are given in introspection, then mental objects — according to formula (42) — are abstracts. The hypothesis that there are also mental concretes (for example, «minds»), is much more doubtful than the hypothesis that some objects are physical concretes.

3.17

There appears here the need, and the opportunity, to distinguish the empirical from the fictitious.

I assume:

(44) For every x [x is empirical *iff* for a certain y (y is observable and x acts on y).

(45) For every x [x is fictitious *iff* it is not the case that for a certain y (y is observable and x acts on y)].

Sometimes formula (44) is given as an explication of the term "reality" [Moore 1953, p. 229].

After it is further specified what an action is, formula (44) does not stand in the qay of assuming that empirical objects include not only some observable objects but also some noumenal objects (*viz.* the *designata* of theoretical terms). After all, we can impose upon empirical objects, for instance, the condition of being in the part-complex/whole relation to them, instead of the condition of acting on the observable objects.

3.18

I assume:

(46) For every x {x is objective *iff* for a certain P[either Px or for a certain y (y thinks that Px) but not both]}.

(47) For every x {x is subjective *iff* for every P [Px *iff* for a certain y (y thinks that Px)]}.

One can (partially) «depsychologise» formulae (46) and (47) and express them as follows:

(48) For every x {x is objective *iff* for every P [Px or for a certain y (y describes the fact that Px) but not both]}.

(49) For every x {x is subjective *iff* for every P [Px *iff* for a certain y (y describes that Px)]}.

Here *y*-s are not only people but also sentences and definitions in particular.

Formulae (46) and (48) will be probably more intuitive if the right-hand side of the equivalence is stated in the following manner:

(50) For a certain *P* ⟨*Px* and it is not the case that for a certain *y* {*y* thinks (*resp.* describes the fact) that *Px*; or it is not the case that *Px* and for a certain *y* [*y* thinks (*resp.* describes the fact) that *Px*]}⟩.

It is now easier to note the intention of this formula, according to which something "exists OBJECTIVELY, i.e. independently of a cognitive subject" [Augustynek 1975, p. 18] or that its "existence is not necessarily conditioned by a thought" [Ajdukiewicz 1923, p. 99].

Subjectivity is correlated, of course, with heteronomy, but they are not identical, contrary to those scholars who "regard matter as self-subsistent" [Russell 1959, p. 51]. Moreover, one remember that in such cases one often speaks about reality and ideality [Kotarbiński 1929, p. 341; Ajdukiewicz 1949, p. 79] or reality and fiction [Moore 1953, p. 211] instead of objectivity and subjectivity.

Le us call all objective objects simply "objects", and all subjective objects "quasi-objects".

First, distinction between concreteness and abstractness and the distinction between extra-mentality and mentality are applicable, strictly speaking, only to objects: quasi-objects are at most quasi-concrete or quasi-abstract, and quasi-extra-mental or quasi-mental. Second, only objects can be observable or material, although some of them are probably noumenal or ideal. Third, all objects are empirical or individual. Fourth, all quasi-objects are noumenal or ideal. Thus we cannot claim that the differences between observability and noumenality are not "ontologically essential" [Czarnocka 1986: 119,121]. On the other hand, it is true that ontological forms are not identical with epistemological forms. "Objects perceived in different ways need [not] belong to different ontological categories" [Czarnocka 1986: 124]. Fifth, only quasi-objects can be fictitious or universal, though some of them are probably empirical or individual. Thus, since only (individual or universal) fictions are incompatible, only quasi-objects possess the property of incompatibility.

3.19

It remains only to consider the question of whether concretes and abstracts exist, or more generally whether objects distinguished in formulae (1)—(5) exist.

I am suspicious of the view that existence is not a property since it is not backed by adequate arguments. An answer to the question "Which objects exist?" should be preceded by an answer to the question "Which intuitions ought to be preserved?". It seems to me that, for example, the following statement comes closest to common sense:

(51) For every *x* (*x* exists *iff* *x* is objective).

Existence would not be a property only if it had could not be identified with any property from among the properties characterised in this paper. But then the question of what exists would be *questionae gustuum* and not *questionae facti*.

3.20

The problem of ontic forms puts us to a great deal of trouble not so much because scholars differ as to the accepted solutions as because we do not exactly know what these differences consist in.[6]

---

[6] For detailed discussion of these problems cf. my "Ontological minimum" [Augustynek and Jadacki 1993].

4. ON GOOD, NECESSITY, AND SUFFICIENCY

Below, I would like to explicate the notion of good. I do not intend to reconstruct all the semantic subtleties of the «ordinary» uses of this word which can be perceived by an experienced linguist (and of words related formatively and semantically). I want only to propose some terminological regulations well suited for setting some axiological standpoints in order, which will make it possible to present these standpoints as definite ideological choices and not as trivial (*scil.* analytical) consequences of the accepted definitions.

4.1

Let us compare the following dependencies:

(1) It is always the case that if one has friends, one escapes trouble.

(2) It is always the case that if one does not have friends, one does not escape trouble.

(3) Often, if one has friends, one escapes trouble.

(4) Often, if one does not have friends, one does not escape trouble.

Having friends, according to:

(1) is a SUFFICIENT condition of escaping trouble (because with friends, one always escapes trouble);

(2) is a NECESSARY condition of escaping trouble (because without friends, one never escapes trouble);

(3) and (4) is a CONDUCIVE condition of escaping trouble (because with friends, one often escapes trouble, and without friends, one often does not escape it).

Contingent on which of the dependencies occurs, different strategies of behaviour are rational.

In case (1) holds, it is rational to behave according the recipe:

(5) To escape trouble, it is SUFFICIENT to have friends.

This recipe can be schematically expressed in the following way:

(6) If x wants to escape trouble then it is sufficient for x to have friends,

and more generally:

(7) If x wants α, then it is sufficient for x to β.

In case (2) holds, it is rational to behave according to the recipe:

(8) To escape trouble, it is necessary to have friends.

Schematically:

(9) If x wants to escape trouble then it is necessary for x to have friends,

and more generally:

(10) If x wants to α then it is necessary for x to β.

In case (3) and (4) hold, it is rational to behave according to the recipe:

(11) To escape trouble, it is GOOD to have friends.

Schematically:

(12) If x wants to escape trouble then it is good for x to have friends,

and more generally:

(13) If x wants to α then it is good for x to β.

N.B. Henryk Elzenberg always said "It is good THAT x has friends" [Wolniewicz 1986, p. 81]. I do not see good reasons for such a limitation. On the other hand, formulae (11)—(13) go well with Marian Przełęcki's suggestions that it is more cautious to use comparative predicates of evaluation [Przełęcki 1996, p. 34]. In fact, to escape trouble, it is BETTER to have friends than not to have them; because probably (*cf.* formulae (3)—(4)) one more often escapes trouble if one has friends than if one does not have them.

In practice, the content of formula (11) is expressed in various ways, for instance by means of the formulae:

(14) Having friends IS GOOD FOR escaping trouble,

(15) Having friends facilitates escaping trouble,

or shortly:

(16) It is good to have friends in case of trouble.

Schematically:

(17) *B* is good for *A*,

or

(18) *B* facilitates *A*.

However, I do not think that the formula:

(19) *B* is good for *A* always and only if[1] *B* facilitates *A*

can be treated as a theoretically fruitful definition of formula (17). It is useful only for a person who (intuitively) understands phrase (18) better than phrase (17).

Formula (17) shows more clearly than, for example, formula (13) that "good" signifies a certain relation.

4.2

As can be seen, formulae (5), (6) and (7) speak about sufficiency; formulae (8), (9) and (10) — about necessity; and formulae (11), (12) and (13) — about good. These formulae show clearly the relation between sufficiency, necessity and good. Sufficiency is a sufficient condition of something, necessity — a necessary condition, and good — a conducive condition. Thus, we have:

---

[1] I consequently read the functor of equivalence: "always and only if", because the widespread formula "only and only if" is misleading.

(20) If it is sufficient (for something) to have friends then it is good (for the same thing) to have friends.

(21) If it is necessary (for something) to have friends then it is good (for the same thing) to have friends.

In the case of (21), we can say shortly: it is always good if it is as it should be. If the reconstruction of Elzenberg's views made by Bogusław Wolniewicz is correct then according to Elzenberg it is good only if it is as it should be [Wolniewicz 1986, p. 80].

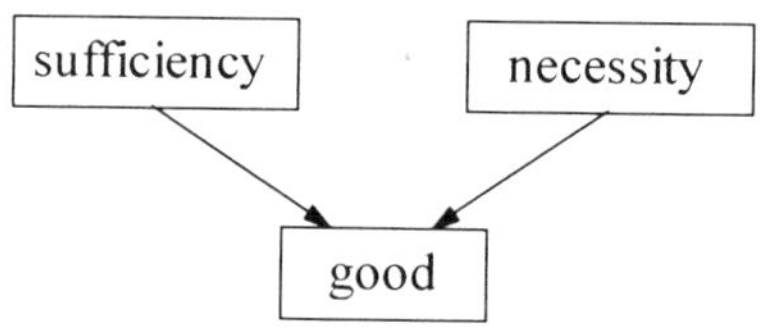

Fig. 3. Dependencies between good, sufficiency, and necessity

It is worth noticing that neither sufficiency entails necessity (because sufficient conditions may not be necessary), nor necessity entails sufficiency (because necessary conditions may not be sufficient). Of course, good entails neither sufficiency, nor necessity (because conducive conditions may be neither sufficient nor necessary).

4.3

Let us «cut off» the antecedents of formulae (5)—(13). We shall obtain respectively:

(22) It is SUFFICIENT to have friends.

(23) It is sufficient for *x* to have friends.

(24) It is sufficient for *x* to β.

(25) It is NECESSARY to have friends.

(26) It is necessary for *x* to have friends.

(27) It is necessary for x to β.

(28) It is GOOD to have friends.

(29) It is good for x to have friends.

(30) It is good for x to β.

How should we understand these unrelativised sufficiencies, necessities and goods?

We can treat formulae (22)—(30) as shorter versions of formulae (5)—(13). If someone says that it is good to have friends, s/he has in mind that it is good to have friends to certain end (e.g. to escape trouble).

We can also treat formulae (22)—(30) as formulae expressing beliefs that there is something which is sufficiency in general, necessity in general or good in general. The expression "in general" may mean here the same as "to every end" or as "*simpliciter*".

When one talks about universal or absolute goods (*resp.* sufficiencies or necessities), one can have in mind just such goods (*resp.* sufficiencies or necessities) in general: something that is good (*resp.* sufficient or necessary) to every end or irrespective of any end. If having friends is a universal good then it is good to have friends to every end or irrespective of any end.

Let us agree that sufficiencies, necessities and goods are certain values. The controversy between absolutists and relativists concerns just the universal absolute values: whether they exist (as absolutists claim) or not (as relativists claim).

How could one settle this controversy? In the case of goods, the absolutist should indicate that at least one good is good to every end or irrespective of any end; while the relativist should argue that no good satisfies this condition.

On the other hand, it also happens that the ends themselves are goods (or sufficiencies, or necessities, respectively); but it only happens! Thus, is it good to want the realisation of such ends? Is it good, for example, to want to escape trouble?

The relativist will answer: sometimes (in view of certain ends) it is good, but sometimes (in view of other ends) it is not good. For instance, often if one WANTS to escape trouble, one escapes them. Thus, in order to escape trouble, it is good to want to escape it. However, such a dependence between the want of escaping trouble and, for example, longevity does not exist — even if escaping trouble favours longevity.

The absolutist is expected to do something more: to decide whether ends can be universal or absolute goods.

4.4

What is the controversy between objectivists and subjectivists about? Sometimes, it is said that it concerns the problem of whether values (sufficiencies, necessities and goods, in particular) depend on our caprices or not. Does it happen, for example, that something is good irrespective of the fact that somebody considers it to be good?

Let us notice that both formulae (20) and (21) can be expressed in terms of beliefs. The latter formula would have the following shape:

(31) If x thinks that it is necessary to have friends then x thinks that it is good to have friends.

This is not subjectivism yet. Only someone who «cuts off» the second "x thinks that" in formula (31) counts as a subjectivist, i.e. someone according to whom:

(32) If x thinks that it is necessary that the fact that p occurs then the fact that p is good.

We can consider volitionalism as a vcariant of such a subjectivism. The volitionalists assume additionally that:

(33) x thinks that it is necessary that the fact that p occurs ALWAYS AND ONLY IF x wants the fact that p to occur.

From assumptions (32) and (33) we can derive:

(34) If x wants the fact that p to occur then the fact that p is good.

It seems to me that the controversy under discussion is in fact about occurrence of dependencies such as (1)—(4), which are at the base of rational behaviour. Whether these dependencies occur or not is certainly not a matter of anyone's caprice. Either EVERY person who has friends escapes trouble, or NOT EVERY; either EVERY person who does not have friends does not escape trouble, or not every; either SOME (MANY?) persons who have friends escape trouble (and SOME (MANY?) persons who do not have friends, do not escape trouble), or NONE (NOT MANY?) of them. It is worth noting that here the facts are such that having friends is neither sufficient (*scil.* a sufficient condition), nor necessary (*scil.* a necessary condition) to escape trouble; but it is good (*scil.* a conducive condition) to this end (and, one might add, not only to this end — probably to any end).

However, many people would like to know something else: whether it is good *simpliciter* to have friends — irrespective of any end.

This question can be answered only by a person who has her/his own intuition about values.

What can somebody who is NEITHER a cognitivist NOR an intuitionist consider to be an analogue of formulae (3)—(4) in the case of the ethical declaration: "Morally good acts are altruistic acts, *ergo* acts motivated by the concern for other persons' good" [Przełęcki 1996, p. 37]? Perhaps s/he should consider the formula: Often, if someone acts altruistically, s/he lives happily. Thus: in order to live happily, it is good to care for other persons' good. Now, let us note that someone's saying "It is good to live happily" would make sense only when happy life was a conducive condition for a certain end. Incidentally, on such an approach, it becomes absolutely clear that the altruist must know form another source what her/his protégé's aims are. In other words, the altruist must know what her/his protégé really WANTS because only then will s/he be able to look for the conditions which favour these ends. Of course, it is easy to project one's own aims.

5. ON ESSENCE

5.1

My view on essence are indebted to Rev. Jan Salamucha in two ways. I appeal to his method of inquiry as well as to the results acquired by means of this method. First, like Salamucha, I use the tools of logical analysis because I believe that only these tools make it possible to understand the problem precisely and present it clearly. Second, I profit directly from Salamucha's conclusions contained in his paper "Z historii jednego wyrazu ("istota")" [From the History of One Word ("Essence")], published posthumously in 1946 [Salamucha 1946].

In order to analyse the notions of "essence" it is not enough to use exclusively either the language of the set theory or the language of the first-order predicate calculus. Hence I join these languages together, modify and enrich them. Since I place great emphasis on the manner in which formulae are read natural language, I want it clearly understood that:

(a) the expression of the form "Px" ought to be read as "x has the property P" (but not as "x is P");

(b) the expression of the form "t: Px" ought to be read as "at moment t: x has the property P";

(c) the expression of the form "$x \in Z$" ought to be read as "x is Z" (or as: "x belongs to the set of Z-s").

The symbol "$C/\supset$" indicates that in a given case it might be implication "C", implication "$\supset$", as well as equivalence "E".

5.2

There are three inessential notions of "essence". What I have in mind here are notions fully reducible to the notion of "property" (or possibly "«necessary» property" or "peculiarity"), "invariable property" or "a sum of properties".

Let us consider the following definition:

(1) $(x)(I)\, E\, (I \text{ is the essence of } x)\, (Ix)$.

This definition simply identifies the essence of a given object with any of its properties. We probably will not find such a definition in the writings of any philosopher. However, let us consider the following formula:

(2) When we speak of «what» anything is [*scil.* its essence], we mean what it is of itself (τό τί ἦν εἶναι ἑκάστον ὁ λέγεται καθ'αὑτό) [Aristotle –323, p. 135].

It seems to me that the idea of formula (2) can be reconstructed as follows:

(3) $(x)(I)\, E\, (I \text{ is the essence of } x)\, (y)\, C\, (x = y)\, (Iy)$.

Let us notice that:

(4) $(x)(I)\, C\, (I \text{ is the essence of } x)\, (Iy)$

and that:

(5) $(x)(P)\, E\, (Px)\, (y)\, C\, (x = y)\, (Py)$.

In virtue of (4) and (5) every property of a given object is its essence. Thus, the definition (3) does not allow to distinguish essential properties from inessential ones.

We can avoid this consequences in two ways.

Firstly, we can introduce the functor of necessity to the *definiens* of the definition (3). In such a case the essence of a given object would be identified with its «necessary» property:

(6) $(x)(I)\, E\, (I \text{ is the essence of } x)\, (y)\, C\, (x = y)\, \Gamma\, (Iy)$.

The following formula goes in this direction:

(7) Grasping the essence of a given thing, we grasp its constitutive properties; deprived of these properties, this thing cannot be what it is (namely this thing) [Krąpiec *et al.* 1992, p. 103].

Secondly, the consequence mentioned above can be avoided if we recognise as the essence of a given object the peculiarity of this object, i.e. a property, possessing of which is, at the same time, the necessary and sufficient condition of identity of a given object.

(8) $(x)(I)\ E\ (I$ is the essence of $x)\ (y)\ E\ (x = y)\ (Iy)$.

This idea is contained, for instance, in the formula:

(9) The essence is something essential: something in virtue of which an object is what it is; deprived of its essence, an object would not be what it is (namely this object) [Stępień 1964, p. 223].

Now, let us consider the formula:

(10) Aristotle taught that every thing has properties of two kinds. Properties of the first kind are essential to it; they constantly belong to it and result in this thing being exactly this one and not something different. These properties constitute the essence of this thing. Properties of the second kind are variable, accidental, inessential [Czeżowski 1938, p. 41].

In this formula, the definition (3) is enriched by temporal characteristics:

(11) $(x)(I)\ E\ (I$ is the essence of $x)\ (y)\ C\ (x = y)\ (t)\ C\ (t\text{: } y$ exists$)\ (Iy)$.

The definitions considered above do not decide whether essences are particular «simple» properties of a given object (then this object could possess many essences), or whether they can — or should — be properties which are «sums» of all particular properties. Other decisions have been made as well:

(12) The essence of a given object [as a transcendental correlate of its existence — the content of being — is] a complex of all qualifications possessed by this object [Stępień 1964, p. 93—94].

We shall reconstruct this formula in the following way:

(13) $(x)(I)\ E\ (I$ is the essence of $x)\ (P)\ E\ (Px)\ (P \in I)$,

allowing that «existence» is not a property. Thus, according to this approach, the essence of a given object is the so-called full content, that is the sum of properties of this object.

5.3

Rev. Salamucha has noticed that the notion of "essence" discussed above refers to essences of individual objects (or individuals). The difficulties with such an understanding of "essence" have been known by the scholastics who tried to eliminate them by linking essences, like Plato, not with individuals but with species (or ideas). For instance, one of the scholastic definitions of "essence" runs as follows:

(14) The essence of a given object is that without which this object does not belong to a given group ([a given] species) of objects [Salamucha 1946, p. 72].

We find similar formulae also elsewhere:

(15) The person uttering the sentence "Quadrilaterality is an essential feature of a square" has in mind that the expression "quadrilateral" can be predicated of the expression "square" analyticallly, or, in other words, s/he has in mind that the sentence "Square is quadrilateral" is analytical. Analytical sentences are understood here as sentences derived from axioms of a given language according to the rules of inference (transformation) peculiar to this language [Ajdukiewicz 1946, pp. 17-18].

(16) The general essence of a given object [is everything] that decides about its specific or generic membership, for example, that which constitutes John's humanity, his animality and his rationality [Stępień 1964, p. 223].

The reconstruction of such a notion of "essence" may look as follows:

(17) $(x)(I)$ (Z) E (I is the essence of x as Z) $C/\supset$ $(x \in Z)$ (Iy).

In such case, the essence of Z-s would be their «co-property» (that is a property of all Z-s; e.g. quadrilaterality in the case of rectangles), singularity (that is a property of only Z-s, for example, squareness in the case of rectangles), or — what is the most frequent solution — peculiarity, that is a property common as well as singular to Z-s (for example, planenness, quadrilaterality, and rectangularity taken together for rectangles).

It is worthwhile considering a few interesting cases.

(18) Z is a heterogeneous set (for instance, the set: {Lublin, Stalin, basket of raspberries}).

Of course, such a case can occur only when it is not excluded in advance that the «group», with respect to which the essence is determined, can be any set (provided that it has more than one element). It is quite possible that sets like the one mentioned in (18) have NO essential essence-«co-property» (except for being-something).

(19) Z is a unit set (for instance, the set {John}).

Here the essence-peculiarity is a property possessed ONLY by the unique element of such a set. In such a case, some people talk about the individual essence of a given object:

(20) The individual essence of a given object [is everything] that constitutes «john-ity» of John [Stępień 1964, p. 223].

(21) Z is the so-called natural proximate species (*genus proximum*).

Rev. Salamucha has shown that this natural proximate species could be understood in a way which leads to *CIRCULUS VITIOSUS*.

Let:

(22) (Z) E (Z is a lowermost natural species) (Z_1) (Z_2) C K $(Z_1 \subset Z)$ $(Z_2 \subset Z)$ N (Z_1 differs essentially from Z_2).

Now, let:

(23) (Z_1) (Z_2) E (Z_1 differs essentially from Z_2) N (I_1) (I_2) K $(I_1 \neq I_2)$ (x) K (I_1 is the essence of x as Z_1) (I_2 is the essence of x as Z_2).

Thus, after two steps in defining, the defined notion returns:

(24) I is identical with Z-ness.

The statement that the essence of Z-s (for example, squares) is Z-ness (here: squareness) is a cognitive TRUISM. This interpretation of the notion of "essence" was given by the «enemies» of this notion, i.a. Thomas Hobbes and his followers.

(25) If I differs form Z-ness, then the following cases are possible:
- (a) I is a «superficial» (striking, apparent) property or, on the contrary, a «deep» (overt, invisible) property;
- (b) I is an INVARIABLE (constant) property.

(26) The thesis "(x) C (Ix) $(x \in Z)$" is a LAW of:
- (a) science, which states in particular a causal-effectual dependence (I is a cause of Z-ness);
- (b) language (or is a terminological convention).

Here are examples of two instances of (26):

(27) To know what it is [*scil.* the nature (τό τί ἐστιν) of a thing] is the same as ⟨to know⟩ why it is [the cause of existence of this thing] [Aristotle –347, p. 54].

(28) The essence of a thing is what is indicated by the definition [of this thing]. [...] Thanks to the essence we can say what is [...] [a given object] (*Essentia enim uniuscuiusque rei est illud quod significat definitio eius.* [...] *Per essentiam uero suam quodlibet dictur quid sit*) [Aquinas 1265, pp. 85—86].

5.4

Let us suppose that the only notions of "essence" in use are fully reduced either to the notions of "property", "«necessary» property", "invariable property" and "a sum of properties", or to the notions of "co-property", "singularity" and "peculiarity". In such a case, the introduction of the term "essence" to the philosophical vocabulary would be rationally groundless, for it violates the principle of mono-denomination, which every good terminology [Jadacka 1976, pp. 29—45] constrains. Fortunately, there are two «essential» notions of "essence" and it is the proof of its terminologicality. I shall distinguish one from another by using the term "simple essence" and "complex essence" respectively.

(29) (x) (I) (Z) E (Z is the simple essence of x as Z) K E $(x \in Z)$ (Ix) (y) (P_W) C C $(y \in Z)$ (P_Wy) Γ C (Iy) (P_Wy).

As we see, definition (29) requires that property P be of kind K, i.e. it belongs to a certain distinguished set W. There are two interesting cases here.

The following passage can be considered to be an interpretation of the definition (29):

(30) When we give a definition of some kind of body, e.g. some chemical element, we describe it by giving some group of characteristics which are common to all individuals of that kind and which distinguish them from all other objects. In doing this, we do not mention all characteristics common to individuals of the defined kind, only some of them. [...] Thus, if the characteristics common to [the] individuals of a particular kind, which are mentioned in the definition of that kind, are so chosen that they not only distinguish individuals of that kind from remaining objects but also allow us to infer from them, according to the laws of nature, other important characteristics common to the individuals of that kind, we can say that incorporated in the definition was not only a unique characterisation of the kind, but also an essential characterisation. We say of such a definition that it gives the essence of the defined kind, and the characteristics mentioned in such a definition we call essential characteristics [Ajdukiewicz 1956a, p. 241].

Unfortunately, it is not precisely said what is the IMPORTANCE of the characteristics which are «implied» by a given (simple) essence.

Properties of the kind *W* can be just peculiarities.

(31) To inquire after essence would be to find this properties peculiarly common to all elements of a given range, which indicated the sum of their peculiarly common properties" [Kotarbiński 1956, p. 490].

It is surprising that the condition of implying ALL peculiarities is not always imposed upon the essence (on such an approach).

(32) By properties which are «essential» to the elements of a given extension, we often mean those properties which enable us to find out comparatively the greatest number of important points, specific to the extension of the term in question [Kotarbiński 1929, p. 32].

By the way, the essence can be considered in both cases, (30) and (31), as an invariable property (cf. (25b)), and the thesis "Γ *C* (*Iy*) ($P_W y$)" can be treated as a law (cf. (26)).

We read, for instance:

(33) *O* we mean the set of the permanent properties of the object *O* which determine the totality of its specific properties. In this interpretation, *O* is always something general, and determination is understood in the sense that the property *A* determines the property *B* if and only if the property *B* is an attribute of *O* because the property *A* is its attribute [Kotarbiński 1956, p. 414].

Now we go on to the complex essence.

We have here:

(34) $(x)\ (I)\ (Z)\ E$ (Z is the complex essence of x as Z) $K\ E\ (x \in Z)\ (Ix)\ N\ (EP)\ (Q)\ C\ [Q = (I - P)]\ (y)\ (y \in Z)\ (Qy)$.

Defined in such a way, the complex essence of *Z*-s creates the constitutive intension of a term whose extension is the set of all *Z*-s.

Let us remember:

(35) *A characteristic intension of a term N*, for a certain meaning of that term, is any set *P* of properties such that every *designatum* of *N* has every property from the set *P* and only the *designata* of *N* have every property form the set *P*. [...] But a characteristic intension of a term may be *pleonastic*, i.e. it may include more properties than it is necessary to define the extension of that term. For instance, the intension of the term "square" consisting of the properties "planeness, quadrilaterality, rectangularity, equilaterality, being inscribable in a circle" would be pleonastic since it would include more properties than it is necessary to describe the extension of the term "square". This means that a certain property, for in-

> stance that of being inscribable in a circle, can be deleted from this intension and yet the poorer intension thus obtained would describe the same extension. That characteristic intension which is not pleonastic is termed *constitutive*. A characteristic intension of a term is constitutive if it describes the extension of that term but would cease to describe it should any property be deleted from it. Those properties included in a pleonastic intension that describes a set of objects whose deletion would yield the constitutive intension of the same set of objects are termed *consequential* with respect to the set of the properties covered by that constitutive intension; in other words, they are properties which result from the latter properties. For instance, the property of being inscribable in a circle is consequential with respect to the following set of properties: planeness, equilaterality, quadrilaterality, rectangularity, i.e., it results from the set of those properties. The constitutive intension does not include any property that would be consequential with respect to the set of the remaining properties; on the contrary, a pleonastic intension always does include at least one such property [Ajdukiewicz 1965b, p. 44].

A more interesting possibility connected with the notion of "complex essence" occurs when the *definiens* of the definition (34) is not fulfilled because of the fact that the left hand side of this definition is true, while the right hand side is false. In this case one constituent of the complex essence would imply the other one. Thus, we would come back to the notion determined in the definition (29): one of the constituents of the complex essence would be in fact a simple essence.

5.5

Scholastics contrasted the notion of "essence" (*essentia*, *nota essentialis*) with the notion of "attribute" (*proprium*) and "accident" (*accidens*). The connections between these notions can be presented in the following formulae:

(36) (x) (I) (P) C (I is the essence of x) E (P is an attribute of x) Γ (y) C (Iy) (Py).

(37) (x) (I) (P) C (I is the essence of x) E (P is an accident of x) Δ (Ey) K (Iy) N (Py).

As Rev. Salamucha has noticed in deductive systems these three notions can be interpreted as follows. The essence can be treated as a definitional feature, mentioned in axioms and definitions of the system. The attribute can be treated as a thetic feature, mentioned in theses of the systems. The accident can be treated as an extra-systemic feature, mentioned in extra-systemic theses, that is in the so-called factual statements.

Interpreted in this way, the essence and the attribute are always related to a given set of axioms. It is commonly known that a given system can be axiomatised in MANY ways. Thus NO feature of a given object which belongs to the domain of an axiomatised theory, can pretend to be the ONLY essence of that object. The question of the uniqueness of the essence of real objects is not decided because we have at the moment no axiomatised theory of such objects. The only axiomatised theories we have at our disposal are theories of ideal objects.

5.6

Some people claim that philosophers are obliged to inquire into the essence of things. I do not think so. As a philosopher I have nothing to say about the essence of squareness (this is the task of geometry) or, for example, of dominicanity (this is the task of appropriste monastic constitutions). My duty is only to have something to say on the matter of the NOTION of "essence".

Zdzisław Augustynek wrote in his book *Natura czasu* [The Nature of Time], which essentially [*sic*!] influenced my philosophical views, that the word "essence" is "often treated in an archaic manner, namely metaphysically". And he added: "There is no satisfactory characterisation of the meaning of this term" [Augustynek 1975, p. 9]. My remarks in this section are intended to be a step towards the disarchaisation of the notion (or rather of the notions) of "essense".

PART II. TRUTH AND NONSENSE

6. ON THE DEFINITION AND CRITERIA OF TRUTH

6.1

The essay of truth written by baron of Verulam in 1597 begins with the following words [Bacon 1597, p. 5]:

"What is truth — said jesting Pilate; and would not stay for answer."

This was, indeed, one of the worst of Pilate's mistakes but thanks to it, we philosophers are able to find creative delight in human research rather than being satisfied with some ready-made definitive solution of the problem given by God.

Usually, in writings devoted to the problem of truth the author's attention is focused on the need to distinguish the nature (definition) of truth from the criteria (characterisation) of something's being true. Among others, Jan Łukasiewicz has given a clear formulation of such a principle [Łukasiewicz 1913, p. 55].

6.2

The expression:

(1) $(x)\ E\ (Px)\ (Qx)$

is to be read: For every x (x has the property P *iff* x has the property Q).

If the expression "Qx" in (1) does not contain the predicate "P" then I shall say that:

(a) (1) is a DEFINITION of the expression "Px";

(b) the state of affairs such that Qx is a NATURE of the state of affairs such that Px.

For instance, the formula:

(2) For every x (x has at the end of the 18th century the right to build church organs in the Kingdom of Denmark *iff* x is a builder of organs in St. Bridget's cathedral in Maribo)

is definition of "having, at the end of the 18th century, the right to build church organs in the Kingdom of Denmark" and the state of affairs such that x is the builder of the organs in St. Bridget's cathedral in Maribo is a nature of the fact that x has, at the end of the 18th century, the right to build church organs in the Kingdom of Denmark. (N.B. This function is satisfied by Daniel Wróblewski.)

Moreover if the expression "Qx" in (1) is, but the expression "Px" is not, an observational term then I shall say that:

(c) the state of affairs such that Qx is an ATTRIBUTE — or, in other words, a diagnostic nature — of the state of affairs such that Px.

For instance, in the formula:

(3) For every x (x is a national sanctuary *iff* x is the place where there are tombs of at least 17 kings),

the state of affairs that x is a place where there are tombs of at least 17 kings is an attribute of the state of affairs such that x is the national sanctuary. (N.B. This function is satisfied by the cathedral in Roskilde and the Wawel cathedral in Cracow.)

The expression:

(4) $(x)\ C\ (Px)\ (Qx)$

is to be read as follows: For every x (if x has the property P then x has the property Q).

In the case of (4) again if the expression "Qx" does not include the predicate "P" — I shall say that:

(a) (4) is a CHARACTERISATION of the state of affairs such that Px and of the state of affairs such that Qx; more specifically it is a DISTINCTION of the state of affairs such that Px and a DETERMINATION of the state of affairs such that Qx;

(b) the state of affairs such that Px is a WARRANT of the state of affairs such that Qx;

(c) the state of affairs such that Qx is a BASIS of the state of affairs such that Px.

For instance, the formula:

(5) For every x (if x is the wife of the Danish prince Magnus, then x is a daughter of the Polish duke, Boleslaus the Wry-Mouthed)

is a characterisation of the state of affairs such that x is the wife of Magnus and that x is a daughter of Boleslaus, i.e. in particular it is a distinction of the state of affairs such that x is the wife of Magnus and a determination of the state of affairs such that x is a daughter of Boleslaus. The state of affairs such that x is the wife of Magnus is, moreover, a warrant of the state of affairs such that x is a daughter of Boleslaus; on the other hand, the latter is a basis of the former. (N.B. This function is satisfied by Rycheza.)

Moreover, if the expression "Px" in (4) is, but the expression "Qx" is not, an observational term then I shall say that:

(d) the state of affairs such that Px is a CRITERION — or, in other words, a diagnostic warrant — of the state of affairs such that Qx.

For instance, in the formula:

(6) For every x (if x goes by the name of Dunin then x or an ancestor of x took part in a military expedition on the Kingdom of Denmark),

going by the name of Dunin is a criterion of the state of affairs such that the bearer of this name, or his ancestor, took part in a military expedition to Denmark (as, for instance, Piotr Włostowic did).

If, however, the reverse holds then I shall say that:

(e) the state of affairs such that Qx is a SYMPTOM — or, in other words, a diagnostic basis — of the state of affairs such that Px.

For instance, in the formula:

(7) For every x (if x is a postman in Copenhagen, x wears red livery),

the wearing of red livery is a symptom of being a postman in Copenhagen.

6.3

I would like to make the following comments as regards the above constructions.

First, I leave open the question of whether one should impose the condition of truth on (1) and (4). Clearly, without such a condition one admits the existence of «improper» definitions and characterisations. If someone were to take exception to the fact that for the state of affairs such that *Px* more than one nature might be adduced (possibly infinitely many?) then s/he would have to make its nature relative to each given equivalence.

Second, in formula (1) "*iff*" is an equivalence functor; in formula (4) "if ... then ..." is a functor of implication. Since equivalence is a symmetrical relation, it is a matter of convention whether (1) constitutes a definition of the expression "*Px*" or rather a definition of the expression "*Qx*" provided that, in (1), neither the expression "*Qx*" contains the predicate "*P*" nor the expression "*Px*" contains the predicate "*Q*". It will equally be a matter of convention whether the state of affairs such that *Qx* is a nature of the state of affairs such that *Px*, or *vice versa*.

Third, in my view a question about a nature of a given object becomes a question of a corresponding definition. It seems to me that a solution of such a problem that involves a reductive definition, or indeed a reductive characterisation, might be quite satisfactory.

Fourth, the definitional formula (1) does not allow one to distinguish coextensive characteristics. Thus, it does not allow one to distinguish the definition of "is true" from a claim that ascribes some property to the property of being true unless the property in question is unique. Neither does (1) allow one to single out the so-called constitutive nature. Sometimes it is held that the state of affairs such that *Qx* is a nature of the state of affairs such that *Px* only when the state of affairs *Qx* is not only unique (i.e. is ascribable to all and only those *x*-s that are *P*) but also essential for any *x* to be *P*, that is, from the state of affairs such that *Qx* all the distinguishing characteristics of such *x*-s may be inferred. Unfortunately, it is not clear what exactly is meant here by "infer". I am, moreover, inclined to think that it is not at all desirable to formulate such «essential» definitions, if it is to be a description from which a maximum number of (in this case, epistemological) theses are to follow.

6.4

Should the expression "*Qx*" appear neither in any thesis of the theory of those objects that satisfy the function "*Px*" nor in any definition belonging to that theory other than a definition of the expression "*Px*" where "*Qx*" appears as the *definiendum*, then one can say that a definition of the form (1) is superfluous with respect to that theory. Within the confines of the theory, it would simply be a case of *ignotum per ignotum*. In such a case, the defined term would also be superfluous, indeed inessential! On the other hand, it is difficult to conceive of a situation where some unique characteristics could be *de facto* attributed to objects of a certain kind independently of other such characteristics; the relevant properties can at most remain beyond the scope or interests of a given theory or as yet be unanalysed.

Moreover, it might be thought that the distinction between a definition showing a nature of something, and a formula giving a criterion for that thing might consist in a difference between (1) (a definition of "*Px*") and:

(8) For every *x* (to establish whether *Qx*, it is sufficient to do such and such).

Formula (8) is meant to give a criterion for *Px*, where "*Px*" is understood as in (1). We should not forget, however, that (8) can take, for example, the form:

(9) For every *x* (to establish that *Qx*, it is sufficient to establish that *Rx*).

Condition (9) is at least equivalent to:

(10) $(x)\ C\ (Rx)\ (Qx)$.

So there seems to be no reason why we should not take (10) as a formula giving a criterion for *Px*, on the strength of (1). By the same token, we can regard formula (1) itself as a formula giving a criterion, in so far it only defines an attribute or diagnostic nature.

Formula (10) would not be sufficient for:

(11) For every *x* (to establish *whether or not Qx*, it is sufficient to establish *whether or not Rx*).

A certain substitution of (11) might well be true for:

(12) $(x)\ E\ N\ (Rx)\ (Qx)$.

I realise, finally, that to count all and only diagnostic warrants as criteria is to make a certain terminological stipulation. On the other hand, the terms "characterisation", "determination", "distinction", "attribute", "basis" and "symptom" were, as a matter of fact, introduced without regard for general usage and may properly be viewed as contrary to one's linguistic intuitions.

6.5

The problem of truth breaks down into three questions: what is its substratum; what is its nature; and what is a criterion of something's being true. I shall try to list and, if possible, elucidate even those solutions to the above questions which are *de facto* facile or indeed foolish, but still appear from time to time. For I still hope that by bringing them to light I may succeed in making sure that no one in future considers such views worthy of serious attention.

As to the problem of substratum, we should be first of all clear on one thing. We are only interested in what might be called "the metalogical concept of truth", that is, the theory based on the ascription of truth to propositions. In this, we follow in the tradition of Alfred Tarski [1969, p. 402].

When we speak of propositions, we may use the word "proposition" in one of the following senses:

(i) the grammatical sense, i.e. "sentence" or "sequence of words";

(ii) the logical sense, i.e. "assertion" or "sentence-meaning";

(iii) the psychological sense, i.e. "belief".

Some philosophers are of the opinion that "is true" in its primary application can be predicated only of one king of thing. According to Plato, for example, it is sentences that may be true or not. In the view of Sextus Empiricus and the Stoics in general, only assertions can be true. And according to Aristotle and perhaps Charles Sanders Peirce, it is beliefs that are true or false (cf. [Chwedeńczuk 1984, p. 177]). However, I think that quite apart from the possible result of any investigation into the status of assertions, beliefs, or sentences, conclusions regarding the application of "is true" to anyone of these sets of entities can be easily reformulated to apply to the others. The same could be said of any conclusions about the application of "is true" to sentence-

-tokens and sentence-types. Only practical considerations ought to steer us into a particular mode of speaking.

Let us note that sometimes "proposition" in its logical sense is understood as the class of all those possible worlds in which the sentence expressing the given proposition is true. In doing so, not only does one presuppose that sentences are truth-bearers, but one also creates problems for the talk of true assertions.

Some writers, e.g. John Dewey, predicate "is true" not of individual propositions but of sequences of propositions which constitute a theory. I think that if one can defend Dewey's definition of "is true" then one could also defend his view of what the truth-bearers are (cf. [Chwedeńczuk 1984, p. 178]).

6.6

What does "is true" mean? There are three fundamental answers to the question of the nature of truth: one we call the "semantic definition", the second, the "syntactic definition", and the third, the "pragmatic definition".

I begin with the discussion of the first definition in its classical interpretation.

According to Aristotle, "to say something that is true", is "to say that that which is, is, and that which is not, is not". It is "to hold that that which is separate, is separate, and that which is connected, is connected". It is "to speak consistently with the facts".

We should note that "is true" designates a relational feature and not — as, for example, Roman Ingarden sometimes says [Ingarden 1931, p. 300] — a two-place relation of truth. Just as "I am married" is not to be expanded as "I am in the relation of being married to a certain person", so equally "This proposition is true" is not to be expanded as "It is in the relation of being true to a certain event". To establish the truth of something is thus not tantamount to establishing that some relation holds (e.g. the relation of concordance or of similarity, as John Langshaw Austin wanted; cf. [Szymura 1974, p. 175]), which relation, moreover, which admits of degrees and passes this feature on to truth. Rather, to establish a correlation between of an event and a given proposition. Establishing what is the *designatum* of a given proposition (as opposed to searching for some kind of similarity) may result in varying degrees of success (of certainty). This degree will be inherited by the (rational) *assent* to this proposition but not by its truth.

Now, I should like to propose one of the following definitions as an explication of these comments:

(13) For every proposition *S* and event *Z* [if *S* refers to *Z* then (*S* is true *iff* *Z* occurs)];

(14) For every proposition *S* [*S* is true *iff* for a certain event *Z* (*S* refers to *Z* and *Z* occurs)].

It is usually assumed that *Z* satisfies the following condition: $Z \neq S$; if we agree that no proposition is an event, that condition may be ignored.

I do not agree with Tarski that a definition using the term "refers to" (in his case "designates") would be less clear or unambiguous than Aristotle's original formulations. It appears to me that definitions (13) and (14) are no less clear, yet definitely better constructed.

Definition (13) does not preclude it that for no true proposition which there be an event, which would occur and to which the proposition would refer. On the other hand, definition (14) does not rule out that for every true proposition there is an event to which it refers but which does not occur. If this second state of affairs is held to be unacceptable, we can strengthen (14) as follows:

(15) For every proposition *S* [*S* is true *iff* for every event *Z* (if *S* refers to *Z* then *Z* occurs)]

Should one feel it desirable to avoid both consequences then one could take an explication of the semantic definition (in this interpretation) to be the conjunction of (13) and (14).

6.7

> It is generally accepted that the truth of a sentence is dependent on the meaning of the expressions used in it, but one can also assume conversely, that the meaning of the expressions depends on the truth of the sentence in which they are contained [Czeżowski 1946, p. 226].

What is at stake here is that the (justified) ascription of truth to a given sound (or inscription) allows one to accept that sound (or inscription) as an expression that possesses a given meaning (*scil.* as a meaningful expression) and at the very least to exclude certain of the possible interpretations of its constituents. Equally, the semantic definition — at least in the form of (13) — may be clearly taken as a definition of "is true" as well as of "occurs", or as a formula showing a basis of reference.

It is clear that if we take definition (13) to be a formula that shows a basis of reference, then, in order for it to be diagnostic, we must have at our disposal the criteria of truth and occurrence. If, on the other hand, we take it as a definition of "occurs", then necessarily we cannot do without the criteria of truth and reference.

Continuing this line of thought, one can say that (13) gives directly the meaning of, for example, "proposition" or "event". In the first case, if we have reason to acknowledge that a certain entity is true, in so far as it refers to a given event which occurs, then that entity is a proposition. In the second case, we have to do with what, for example, Bogusław Wolniewicz refers to by saying that "every fragment of reality, that verifies a sentence [i.e. warrants its truth], we call a "situation"" [Wolniewicz 1978, p. 27], or — in our terminology — an "event".

6.8

Both definitions (13) and (14) state no more concerning the nature of propositions than that they are certain objects that refer to some events. I also hold that we do not need to demand that *S* asserts "the existence of something" or "the ascription of a given property to something", even though every proposition that "has a claim to being true" will *de facto* satisfy this requirement (for each one can be rewritten in such a form by means of relevant transformations). General theses would, in this last case, take on the form of propositions asserting the relevant dependence; propositions of the form "*p* or it is not the case that *p*" would assert the occurrence of at least one of the relevant events, etc.

The application of the above definitions demands, of course, a satisfactory solution to the problem of the semantic correlates of propositions which are (logically) complex, i.e. we need an ontology of situations.

One could characterise the ontic status of events — the correlates of propositions — as "that which can be represented", "that which is possible", "that which can be designated", where false propositions also refer to such possible (representable) events, only in such cases the events do not occur. The expression "Z occurs" cannot therefore be simply substituted by the expression:

(16) $(Ex)(x = Z)$.

It would be quite permissible to speak of a real and fictional occurrence etc.; the sense of the quantifier "(Ex)" — i.e. "in the (suitably circumscribed) domain, there is an object x such that ..." — does not permit such qualifications. Although the fact that S refers to Z is true implies formula (16), it does not imply that Z occurs, for instance, in the real world. Obviously, if the kind of substitution we have been talking about were allowed, it would follow from (13) that every proposition referring to some event is true merely in virtue of its so referring.

Just as in the case of the terms "characterisation", "determinations", etc., the use of the word "event" in a certain context might upset certain linguistic intuitions. This is the case, for example, when one applies it to the *designatum* of the proposition:

(17) $(2 + 2) = 4$,

or

(18) From the beginning of the 16th century every king of Denmark whose Christian name was Frederick was the successor of a king of Denmark whose Christian name was Christian.

The term "event" has, however, certain virtues (of a notational nature) which offset such shortcomings.

The matter of when we actually have to do with the occurrence of a certain event is a question of the diagnostic, or, more generally, epistemological conditions of its occurrence. The semantic definition is neutral as regards the debate concerning those criteria.

6.9

If one regards as acceptable that description of "denoting" (or, better, "designating") for names, which is based on the sense of the *epsilon* in Leśniewski's ontology then a similar explication of "referring" (or "signifying") in definitions (13) and (14) will be equally instructive, or mysterious, as the case may be! Consider names, for example:

(19) "Daddy" designates a father *iff* for every *x* (*x* is a daddy *iff* *x* is also a father)

Let us assign the symbol "*d*(...)" to whatever occurs in (19) as a name after "is", and the symbol "*k*(...)" to whatever occurs as a name after "designates". On this basis (19) becomes:

(20) "Daddy" designates *k*(father) *iff* for every *x* [*x* is *d*(daddy) *iff* *x* is also *d*(father)].

Clearly (20) is equivalent to:

(21) "Daddy" designates *k*(father) *iff* for every *x* [*x* is *d*(*designatum* of the name "daddy") *iff* *x* is also *d*(*designatum* of the name "father")].

Let us suppose that we cannot name a given name by means of quotation marks, but we know of this name that it was the first name uttered by my daughter. Then we can say:

(22) The first name uttered by my daughter designates *k*(father) *iff* for every *x* [*x* is *d*(*designatum* of the first name uttered by my daughter) *iff* *x* is *d*(father)].

We do not, however, know how to form theses that would be completely analogical to (20) because we have no idea what to put in the brackets in the left hand side of the equivalence after the quantifier. Things are exactly the same with the expression "refers to" for propositions. Here we have:

(23) "It rains cats and dogs" refers to the fact that it rains cats and dogs *iff* for every *x* [*x* is *d*(it rains cats and dogs) *iff* *x* is *d*(the fact that it rains cats and dogs)],

where we can read the expression in the square brackets, as for example:

(24) *x* is raining cats and dogs *iff x* is the fact that it is rains cats and dogs.

If however, we do not have the relevant terms in quotes, then we are left with the formula:

(25) The proposition *S* refers to the event *Z iff* for every *x* [*x* is *d*(*designatum* of the proposition *S*) *iff Z* is *d*(*designatum* of the name of the event *Z*)].

Let us notice that the first constituent of formula (19) is usually given in the form:

(26) "Daddy" denotes the set of fathers.

In the case of formula (25), the problem of verifying the identity of events immediately arises; but the question of a criterion of identity appears to me to be no more «mysterious» than in the case of things.

6.10

We must add that the congruity of semantic function of propositions and names may also appear in this parallel: the truth of a proposition and the non-emptiness of a name. Just as the proposition:

(27) The building has a colonnade of 80 pillars

is true *iff* it occurs — for instance, in reality — that the building has a colonnade of 80 pillars (as in the case of the building of the National Museum in Copenhagen), so, for example, the concept:

(28) Harold Bluetooth's son

is non-empty *iff* there is — e.g. in reality — a (certain) Harold Bluetooth's son (for instance, Sweyn).

The existence of the parallel referred to above does not as yet decide the congruity of the semantic function of propositions and names. There is also a possibility that both propositions and names fulfil the same semantic function of referring to certain objects but the semantic function of propositions does not end there; the question of the consequences of the nominalisation of propositions may indeed be replaced by the question of the consequences (and indeed, very admissibility) of the «sententialisation» of names.

We should note that Tarski would certainly impose a stronger condition on the relation of referring, namely the condition of synonymy of the proposition *S* and the name of the event *Z* (cf. (35), below). Personally, I would be inclined to impose such restrictions on the equivalence as to render the concept of "synonymy" superfluous. It is worth noting already at this point that when Tarski sets out his lists of (true) atomic formulae, he *de facto* invokes, perhaps necessarily, intuitions connected with the term "reference". It is irrelevant that he undertakes this ordering not directly for those formulae, but for their constituents (individual variables and predicates).

If we are talking of the correlation of some objects, we can mean correlation with respect to a relation which is either defined or open (in this latter case, it will be acceptable to say of any two objects that they are correlated with each other). Reference is therefore not some arbitrary correlation. For when we carry out this kind of correlating, we are committing ourselves to accepting the truth of a given proposition in certain circumstances.

Sometimes it is required that reference satisfies the following condition:

(29) If *S* does not refer to *Z* then *S* refers to not-*Z*.

Neither (13) nor (14) give any basis for such an understanding of the term "reference".

6.11

Let us stress now that neither are definitions (13) and (14) logically equivalent nor does one imply the other.

It is very important, that only the definition (13) does not lead to the antinomy of the liar, at least in so far as this would resemble Łukasiewicz's version (cf. [Tarski 1933, pp. 157—158]). Using "*Z*" to stand for:

(30) *Z* is not a true sentence,

by (13) we obtain:

(31) If "*Z* is not a true sentence" refers to the fact that *Z* is not a true sentence then "*Z* is not a true sentence" is a true sentence *iff* it is a fact that *Z* is not a true sentence.

In accordance with our initial convention, we can substitute "*Z*" for "*Z* is not a true sentence". Then we get:

(32) If *Z* refers to the fact that *Z* is not a true sentence then *Z* is a true sentence *iff* it is a fact that *Z* is not a true sentence.

Things are otherwise in the case of (14). On this basis, we get:

(33) "*Z* is not a true sentence" is a true sentence *iff* "*Z* is not a true sentence" refers to the circumstance that *Z* is not a true sentence — and it is a fact that *Z* is not a true sentence.

Using substitution as before, we obtain:

(34) "If *Z* is a true sentence that it is a fact that *Z* is not a true sentence.

using substitution as before.

6.12

Formula (13) has the virtue of being free from the antinomy of the liar. However, it is not satisfactory, as a definition of "truth".

It is commonly accepted that the so-called convention (T), formulated by Tarski, is the criterion of adequacy of a definition of "truth". According to this convention:

(T) The sentence "Snow is white" is true *iff* snow is white.

Let us substitute in formula (13) the sentence "Snow is white" and the expression "the fact that snow is white" for *S* and *Z* respectively. Let us agree also, that:

(35) The fact that snow is white occurs *iff* snow is white.

After such a substitution, convention (T) does not follow from (13). Thus formula (13) is not an adequate definition of "truth", if the adopted criterion is diagnostic.

But is convention (T) a good criterion of adequacy of a definition of "truth"? I have doubts as to its suitability. To present my doubts in this respect, I shall reformulate convention (T) in the following manner:

(T') For a certain x [x = "Snow is white" and x is true *iff* for a certain z (z = snow and z is white)].

If this reformulation is admissible — and in my opinion it seems to be admissible — then it is possible to object against convention (T) that it is not a good criterion of adequacy of a definition of "truth". It is because of the fact that according to convention (T) the state of affairs consisting in the fact that snow is white should imply the state of affairs consisting in the fact that a certain sentence (here: "Snow is white") is true. However, the latter state of affairs — at least according to some intuitions — should imply the existence of this sentence. Thus, the occurrence of a certain extralinguistic state of affairs should imply the existence of a certain (true) linguistic expression. In fact, however, snow could be white although no language appeared in the world.

Let us note that, on the same intuitions, one can question also the contraposition. If we agree that the formula:

(36) It is not the case that snow is white *iff* for a certain z (z = snow and it is not the case that z is white)

gives an admissible paraphrase of the sentence, being the left left-hand side of this formula then it is not the case that:

(37) If for a certain z (z = snow and it is not the case that z is white), then for a certain x (x ="Snow is white" and it is not the case that x is true).

We should agree that the occurrence of a certain extralinguistic state of affairs implies the existence of a certain (false) linguistic expression.

6.13

The intuitions ignored by formula (T) could be expressed by, what we may call, the convention (J), though it is not strictly speaking a convention but *questio facti*. It reads as follows:

(J) The proposition "Snow is white" is true *iff* the proposition "Snow is white" states the occurring (*scil.* in reality) event consisting in the fact that snow is white.

Convention (J) lacks the simplicity of convention (T). However, is simplicity really *sine qua non* condition of adequacy?

Unfortunately, formula (13) fulfils unfortunately neither convention (T) nor convention (J). It causes also a great deal of trouble from the ontological point of view. Variable "Z" ranges over the class of events which some of which can occur while others can be merely events stated by a certain proposition. Let us call these two subclasses "real events" and "imagined events", respectively. Such a differentiation makes it necessary to indicate what the relation of a given imagined event (stated in a certain proposition) is with respect to the real event («making» this proposition true). The analyses made by phenomenologists show explicitly the troubles with the problem of concordance between analogous events belonging to such different ontic categories. I do not want to decide whether or not the so-called purely intentional events are theoretical fictions. I think, however, that it would be good to give a definition of "truth" using only the notion of "real event".

I hope this condition is fulfilled by the following definition:

(38) For every proposition S [S is true *iff* for a certain event $Z_{\in R}$ (S states Z)],

where R is the class of real (*scil.* occurring) events. To be plain:

A PROPOSITION IS TRUE ALWAYS AND ONLY IF IT STATES A REAL (OCCURRING) EVENT.

Definition (38) is immune to the antinomy of the liar. It also does not lead to the troubles with imagined events and their relation to real events. Furthermore, it has the virtue of fulfilling the convention (J). After a proper substitution (cf. sec. 2), we receive this convention directly from definition (38).

Let us add at once that:

(39) For every proposition S [S is false *iff* it is not the case that for a certain event $Z_{\in R}$ (S states Z)];

(40) For every proposition *S* [the negation of *S* is true *iff* it is not the case that for a certain $Z_{\in R}$ (*S* states *Z*)];

(41) For every proposition *S* [the negation of *S* is false *iff* for a certain $Z_{\in R}$ (*S* states *Z*)].

Such a conception assumes that the set which variable "*S*" ranges over, contains no propositions containing the functor of negation. Is it possible to justify such an assumption?

The one-place functor of negation is recognised in the classical propositional calculus as a functor of the same syntactical status as two--place propositional functors with propositional arguments. The semantic correlates of these functors do not seem to be, however, of equal ontic status. The two-place functors (especially the most important of them, i.e., the functors of conjunction and implication) have some ontic dependencies as their semantic correlates; i.e. the dependencies between events stated by arguments of these functors. On the other hand, the functor of negation is read "It is not true that" not by accident. If so then it is not fully groundless to regard it as a degenerated metalinguistic functor ("The proposition *S* is not true / is false").

6.14

Somebody can reject definition (38) on the ground that it implies the occurrence of negative events, which are ontologically suspicious.

My riposte is this. According to definitions (38) and (39), we have:

(42) For every proposition *S* (the negation of *S* is true *iff* *S* is false);

(43) For every proposition *S* (the negation of *S* is false *iff* *S* is true).

Then according to definition (39), if a proposition is false, it states no real event. So, what does it state?

The answer to this question is twofold. One can assume either that:

(i) a false proposition states NOTHING,

or that

(ii) a false proposition states an imagined (i.e. fictitious) event.

Thus, if one assumes either (42) or (43) then it is not necessary to introduce ontologically suspicious objects in the form of negative events.

Solution (i) generates at once the question of whether false propositions are propositions at all. After all they do not fulfil the function of stating, i.e. the function essential to propositions as propositions. This problem is analogous to the problem, that arises with regard to empty names: they also do not fulfil the function essential for names, i.e. the function of designating. In both cases, one should define "proposition" and "name" — according to such an attitude — independently from their semantic functions, i.e. by using, for example, suitable syntactic or pragmatic properties.

On the other hand, solution (ii) can lead to the risk of the return of the troubles indicated above, if on the ground of the presented definition of "false" we use the notion of imagined events. Fortunately, it does not have to be so because these imagined events are not necessary to define "truth", and the problem of the comparison between them and the «corresponding» real events arises only in the case of the definition of "truth".

I would like at once to warn against expressing formula (38) in such a way:

(44) For every proposition *S* [*S* is true *iff* for a certain event *Z* (*S* states *Z* and *Z* occurs)],

because, as I think, in the case of (44) it is easy to reconstruct the trouble lately indicated for formula (13). One can appeal to the standard practice, according to which:

(45) $(Ey_{\in Z})\ (Py)$

and

(46) $(Ey)\ K\ (y \in Z)\ (Py)$

are different versions of the same proposition. But I would like to point to a certain detail. The equivalence of the form:

(47) *E* (45) (46)

does not hold for every pair of "*P*" and "*Z*". Namely, it does not hold when the class of objects being *P* is a subset of the class *Z*. Proposition (46) can be false if there are no *Z*-s while the intentions connected with propositions like (45) speak for the interpretation according to which proposition (45) is false because no *Z*-s are *P*.

6.15

Now we go on to the formalistic interpretation of the semantic definition of truth. I put it into the following form:

(48) [For every proposition (that a substitution of) "*S*"] "*S*" is true *iff* *S*.

The variables in (13) and (14) are nominal variables; here, however, in (48) we have sentential variables. Therefore, we can allow the placement of:

(49) "*S*" is true *iff S*

within the scope of the quantifiers binding "*S*" only if we accept to the substitutional interpretation of quantification. However, this ultimately leads us to invoke the concept of "truth". So we cannot, after all, apply this interpretation in a formula meant to define that concept itself.

Tarski has drawn our attention to the difficulties inherent in the formulation of the convention (due to the fact that "*S*" is *suppositione materialis* of the sentence and not of the respective letter) and to the inapplicability of this formula to natural (ordinary) language. It should be emphasised that Tarski does not exclude the possibility of constructing a formula which might be satisfactory in this respect.

One might get the impression that formulae (13) and (14) are competing solutions to Tarski's problem. Such an impression would be mistaken, however. Tarski has shown the way in which the truth of complex sentences depends on the truth of simple sentences. At most, (13) and (14) compete with (48), or, in other words, with the «convention» *T*; and Tarski himself has best shown the limitation of the latter. The result that he arrived at, in the form of the thesis that the set of true sentences in a rich language is not definable in that language, is generally thought of as a fundamental breakthrough in contemporary analyses regarding the concept of truth.

6.16

The semantic definition of truth has also the so-called nihilistic interpretation. I express this interpretation in the following words:

(50) "The proposition S is true" means the same as "S".

According to this interpretation (whose proponents include, among others, Rudolf Carnap) the words "is true" are held to be logically superfluous, i.e. a mere stylistic decoration.

Tarski himself has noted that the nihilistic conception of truth excludes the formulation of hypotheses such as, for instance, that all sentences uttered by a given mathematician are true [Tarski 1969, pp. 410—411]. We would be right to reject the synonymy of the *definiens* and *definendum* in (50). The situation alluded to by Tarski testifies to this difference in meaning.

6.17

The second fundamental answer to the question about the nature of truth is the so-called "syntactic definition of truth". It goes like this:

(51) For every proposition S [S is true *iff* for every proposition T (if T is true, then S coheres with T)].

This formula can have two interpretations, depending on our understanding of "coherence".

According to the congruential interpretation:

(52) For every proposition S [S is true *iff* for every proposition T (if T is true then S coheres with T)],

whereas according to the inferential interpretation:

(53) For every proposition S [S is true *iff* for every proposition T (if T is true then S follows from T)].

Kazimierz Twardowski is right in regarding (53) as already carrying the flaw of circularity: for coherent sentences ("sentences in agreement") are necessarily those that may be true together [Twardowski 1975, p. 274].

Under both interpretations (51) is not a definition of "is true", for the predicate "is true" occurs on both sides of the equivalence. To give (51) the form of a definition, we must recast it in one of two versions: either radical or limited, for both the congruential and the inferential interpretation.

The radical version of formula (51) has the following scheme:

(54) For every proposition *S* [*S* is true *iff* for every proposition *T* (*S* coheres with *T*)].

Since, for every proposition we can find another proposition inconsistent with it, we can see straightaway that (54) in the congruential interpretation is unacceptable. We could accept the radical version of (51) in its inferential interpretation — which is simply a definition of logical truth — if we had a criterion of implicational consequence that did not invoke "is true". However, as Tadeusz Kotarbiński has noted, "in order to know that a given statement follows logically (implicationally) from any other statement, we must know beforehand that it is true" [Kotarbiński 1929, p. 113]. The same could be said of equivalence, and, to an even greater extent, of semantic consequence.

Because of the difficulty alluded to above, "is true" can be determined only with regard to a certain delimited set of propositions. This gives us the following limited versions of (51):

(55) For every proposition *S* belonging to a given set *X* [*S* is true *iff* for every proposition *T* belonging to *X* (*S* coheres with *T*)];

(56) For every proposition *S* [*S* is true *iff* for every proposition *T* belonging to a given connected set *X* (*S* coheres with *T*)].

Both these versions can be seen as an explication of the view that "the truth of a sentence depends on whether it is a constituent of a connected whole" [Marciszewski 1970c: 216]. One cannot, indeed, exclude the possibility that some might be inclined to see a nature of truth in the internal coherence (in particular the internal consistency) of a given proposition. But this is not even a symptom of the truth of a proposition.

6.18

Some defend the congruential interpretation of the so-called syntactic definition in its limited version — or rather reject the arguments of its critics — by pointing out that the existence of consistent fairy tales "is not a banal truth but a highly contentious statement" [Wolniewicz 1968, p. 186].

Let us note, first of all, that the anticoherentist argument invokes only the *possibility* of constructing a coherent (consistent) series of untrue propositions and not the *fact* that such a series *has* been constructed.

Secondly, we should add that it is not sufficient to point out that in some novel by George Elliot or Anthony Trollope one can find the proposition that, for instance, in a certain year one month was missing (or that such a proposition follows only from the propositions contained in that novel). We must also show that that there is contained therein the additional proposition that every «novelistic» year has twelve months (or that there is some proposition there, from which this follows). But even if this were to be shown, one could still rid the novel of all such unhappy, because contradictory, fragments. What would remain after such an operation would certainly not be a work by Elliot or Trollope, but it would be "a consistent novel".

Thirdly, it is not true that the operations of all detective institutions are based on the supposition that "fantasy in the end will always show up because of some internal contradiction — just like falsity" (cf. also [Szumilewicz 1966, pp. 93ff]). The person carrying out investigations does not so much hope to ensnare the accused by means of some internal inconsistency in his statements, as to catch her/him finally in stating a proposition, which is false — in the semantic sense of falsity — in the light of the testimony of a reliable witness. Sentencing cannot be based on inconsistencies in testimony. The "sequence of syntactic checks" [Drozdek 1980: 186] is broken, when it is assumed that the statements of the reliable witness correspond to facts.

6.19

The pragmatic definition of truth has three interpretations.

The first of them, called the "instrumentalist interpretation", occurs either in the absolute version or in the relative one.

The absolute version can be stronger or weaker:

(57) For every proposition *S* [*S* is true *iff* for every person *O* (*S* is useful for *O*)];

(58) For every proposition *S* [*S* is true *iff* for a certain person *O* (*S* is useful for *O*)].

When we say that a proposition is true only if it is «useful» or «effective», we may be speaking of truth relative to a given person.

The relative version reads:

(59) For every proposition *S* and person *O* (*S* is true for *O* *iff* *S* is useful for *O*).

In the stronger absolute version (to which we limit ourselves here), we have therefore:

(60) For every proposition *S* {*S* is true *iff* for a certain event *Z* and every person *O* [if (*O* intends to bring about *Z* and *O* acts according to *S*) then *O* will bring about *Z*]}.

In (60) it is assumed that the event, which someone intends to bring about, is to his advantage as an «intended result». From an epistemic point of view it might enable her/him to «discover other truths»; from a non-epistemic point of view, it might enable her/him to arrive at certain hedonistic or moral goods. Somebody's action results in success *iff* this person arrives at a given (advantageous) event or realises it.

I'd like to understand the expression:

(61) *O* acts according to *S*

as:

(62) *O* acts, assuming *S* (to be true),

and therefore:

(63) *O* acts according to the assumption that *S* is true.

I do not deny that it is difficult to state clearly what "acts according to the assumption that" means. Some results obtained by Klemens Szaniawski [1973, pp. 190ff]) may help with the explication of this expression.

In any case (60) must take the form:

(64) For every proposition *S* {*S* is true *iff* for a given event *Z* and every person *O* [if (*O* intends to bring about *Z* and *O* acts ac-

cording to the assumption that *S* is true) then *O* brings about *Z*]}.

We can see straightaway that (64) is not a definition in our sense of the word for the term "is true" appears on both sides of the equivalence. Formula (64) would be a definition, if the term "is true" occurred in phrase (63), e.g. in the sense given by the semantic definition. Whether or not we agree with this remark, George Edward Moore is right in writing that the view that all and only true propositions are useful is obviously false.

The above interpretation of phrase (61) is admissible but I cannot decide if it is the only admissible one. In this connection, the claim that (64) is not a definition is inconclusive. But I would not be inclined to reject it on the basis that the explication of (34):

(65) *O* acts in the belief that *S*

can go through according to Carnap's well-known suggestions, since these do not seem to me at all convincing.

6.20

Sometimes reference to the semantic sense of "is true" is even more obvious that in (51). It is sometimes said that "if a belief «respects» the facts, it is useful *iff* it deserves to be called "true"" [Chwedeńczuk 1984, p. 254].

Thus we have the «practicist» interpretation of the pragmatic definition of truth (in the stronger absolute version):

(66) For every proposition *S* {if *S* is true (semantically) then [*S* is true (in the practicist sense) *iff* for every person *O* (*S* is useful for *O*)]}.

For it is presumably the case that a given belief «respects» the facts if it refers to a certain event that occurs, which (according to the classical interpretation of the semantic definition) happens when the above belief is true (semantically). Under both interpretations, instrumentalist and practicist, the definition of "is true" in the pragmatic sense cannot do without the concept of "is true" in the semantic sense.

In those passages of the pragmatists (e.g. William James) which deal directly with truth, one can find also formulations interpretable as variants of the semantic definition (cf. [Chwedeńczuk 1984, pp. 156ff]). Thus, there arises a metatheoretical objection that, as much as the pragmatists (James including) are opposed to the classical definition of truth, that they presuppose it at the same time. I think that this can be resolved because, in the first case, we can understand "classical version of the semantic definition" as pertaining to that version of the semantic definition where "agreement" means "being a picture of" whereas, in the second case, we can apply (13).

In stating above that some passages of the pragmatists can be interpreted as opting for the classical theory of truth, I do not want to say that the metaphysics of (some) pragmatists does not differ from that accepted by some defenders of the classical theory of truth. But this is a question of the ontic status of the perceptual constituents of experience.

6.21

Just like the instrumentalist and the practicist interpretation of the pragmatic definition of truth so the utilitarian interpretation can be formulated either in an absolute (stronger or weaker) version or in a relative version. Below we consider only the absolute version of this interpretation.

(67) For every proposition *S* and event *Z* {if *S* refers to *Z* then [*S* is true *iff* for every person *O* (*Z* is useful for *O*)]}

Contrary to the instrumentalist and practicist interpretation, (67) not only endows the words "is true" with a *new* meaning, over and above that of the semantic definition, but it seems to be also *independent* in relation to the latter.

We should remember what Leon Chwistek wrote about this definition half a century ago:

> Both theoretical and experimental truth are definitely independent of the problem of usefulness. There exist several mathematical theses whose truth nobody doubts. And yet, they lead a miserable existence in works read by nobody. An assertion of experimental truth may be not only useless but even decidedly dangerous, e.g. in the case of a sleepwalker walking across a roof. However, many fundamentally wrong hypotheses have played a decisive role in the deve-

lopment of science, stimulating thought and leading to new investigations [Chwistek 1930—1931, p. 208].

6.22

I think that the reconstructions undertaken and the comments made above allow us to draw the following conclusions.

First, the so-called syntactic and pragmatic definitions are erroneous, for, depending on their interpretation:

(a) they are not generally definitions *sensu stricto*;

(b) they contain obvious circularity;

(c) they necessarily invoke the semantic definition;

or

(d) they prove to be no more than a variant of the latter definition.

Second, under a certain interpretation the so-called syntactic definition imposes conditions that are satisfiable by no true proposition. The pragmatic definition, in turn, imposes conditions satisfiable only by some true propositions.

Therefore, of the three answers to the question of nature of truth, there remains only one: that which finds expression in the semantic definition. The rest I would consign to the scrapheap of the history of philosophy.

6.23

How do we establish whether a given proposition is true?

If we accept the semantic definition under the classical interpretation, we can answer the question in the following way: it is enough to establish whether the event to which the proposition refers, occurs. Sceptics will add that "to establish" means here "to establish with certainty" and that this is never possible. I hold the limitation imposed by the sceptics to be overly strict: we establish the truth of different propositions in different ways and with varying degrees of certainty. For, *ex definitione*, there is no universal criterion, i.e. one that will allow us to recognise the truth of any proposition whatsoever!

It seems that even something that could aspire to the name of "attribute", "diagnostic nature of truth" would have to be a certain disjunction of characteristics.

Every criterion of truth has been "coined [only] for a given domain of sentences". Moreover, often (for instance, in the case of theoretical hypotheses), we have to make do with an indirect criterion. "The search for an absolute criterion of truth for any kind of sentence concerning facts is a hopeless undertaking" [Czeżowski 1970, p. 25].

6.24

I propose the following explication of the evidential characterisation of truth:

(68) For every proposition *S* and event *Z* {if *S* refers to *Z* then [if for a certain person *O* (*Z* is evident to *O*), then *S* is true]}.

What is evident (what arouses the "irrefutable feeling", what "forces itself" on us) may be either objective or subjective. Thus, we have the perceptual and intuitional interpretation, respectively:

(69) For every proposition *S* and event *Z* {if *S* refers to *Z* then [if for a given person *O* (*Z* is just being perceived by *O*) then *S* is true]};

(70) For every proposition *S* and event *Z* {if *S* refers to *Z* then [if for a certain person *O* (*Z* appears necessary to *O*) then *S* is true]}

The evidential criterion — clarity and distinctness of perception — was formulated *expressis verbis* by René Descartes (cf. [Drozdowicz 1980, p. 243]), but had been known previously to the Stoics (cf. [Rubczyński 1906, p. 309]). This characterisation is sometimes formulated in such a way that what is evident are not (perceived or considered) events, but the propositions themselves.

Taking into account the modifications made above (cf. sec. 12), we would express the evidential criterion in the following manner:

(71) For every proposition *S* and event $Z_{\in R}$ [if (*S* states *Z* and it is obvious that *Z* occurs) then *S* is true].

The criterion of falsehood can be expressed accordingly:

(72) For every proposition *S* (if *S* is intrinsicly contradictory then *S* is false).

6.25

Proponents of the syntactic definition of "is true" sometimes express themselves in such a way that their statements may be seen not as giving a definition but rather a characterisation of "is true" — one that might compete with the evidential characterisation.

The coherential interpretation of that structural characterisation reads:

(73) For every proposition *S* [if for every proposition *T* (if *T* is true then *S* coheres with *T*) then *S* is true].

Characterisation (73) may be given a congruential and a inferential interpretation. Like the syntactic definition and the operational interpretation of the structural characterisation, it can also have a radical and a limited version. In the first case, proposition *T* belongs to the «totality of our knowledge» or at least to the broadest set of propositions possible, where "knowledge" means the set of all propositions held to be true, and not just propositions held to be true that are in fact true. In the second case, the set of initial propositions is constituted in particular by propositions held to be true on the basis of an evidential criterion in the perceptual interpretation.

The thesis that truth admits of degrees which is connected with the coherential characterisation (or definition) of "is true", can be defended only if "degree" means "degree (strength) of belief about the truth of the proposition". The greater the set of propositions taken to be true with which a given proposition coheres, the greater the degree of confidence about its truth.

The charge of infinite regress directed against the coherential criterion applies to all criteria — the pragmatic criterion, as Bertrand Russell has shown, but also to the correspondence criterion. The charge is put as follows: when a certain criterion is applied, the result of the test must be reported by means of a proposition, which in turn ought be tested with respect to truth. To avoid this charge, it is sufficient to distinguishing *asserting* that a given sentence is true from *establishing* that it is true.

Similarly, if Nelson's paradox is correctly constructed then it can be avoided by means of any characterisation of truth.

6.26

According to the operational interpretation of the structural characterisation:

(74) For every proposition *S* [if for every proposition *T* (if *T* coheres with *S*, then *T* is true), then *S* is true].

Formula (74) under the (perceptually) limited inferential interpretation (i.e. limited in such a way that the set to which *T* belongs is the set of perceptual propositions) can be seen, as has long been recognised, as a characterisation providing a criterion for general propositions.

Sometimes pragmatists express themselves in such a way that it is as if they hold usefulness to be a symptom of truth:

(75) For every proposition *S* [if *S* is true then for every (or for a certain) person *O* (*S* is useful for *O*].

On the other hand, the so-called pragmatic criterion of truth is in fact a pragmatic criterion of (the duty) of belief:

(76) For every proposition *S*, event *Z* and person *O* [if (it is useful for *O* that *O* believe that *Z* occurs, and *S* states *Z*) then *O* ought to believe that *S* is true],

where "useful" can mean the same as "rational" (e.g. "sufficiently justified"), "morally advisable", etc.

6.27

Aside from the evidential and structural characterisation also the genetic one is considered by some scholars. It has two interpretations: assential and approbative.

The first interpretation occurs in the universal and scientistic version:

(77) For every proposition *S* [if for every person *O* (*O* assents to *S*) then *S* is true];

(78) For every proposition *S* [if for every researcher *B* (*B* assents to *S*) then *S* is true].

It is easy to show that not all propositions held by all (also among researchers) to be true are true.

Some believe that general agreement is not a criterion but the nature of truth (cf. [Dąmbska 1931, p.15]). Perhaps a certain particular («egocentric») sense of "is true" gets in the way here, a sense in which we predicate "is true" of "someone's assertion when we want to assert that we take up the same position as that expressed by the fact that this assertion was made" [Kmita 1964, pp. 119—120].

The approbative interpretation reads:

(79) For every proposition *S* [if for a certain person *O* (*O* assents to *S*) then *S* is true].

The criterion of authority that one comes across sometimes may be taken to be a variant of the genetic characterisation in its approbative interpretation when the people whose beliefs are the criterion of truth are specified.

Sometimes it is held that someone's assent is not a criterion but a symptom of the truth of the proposition in question. One should bear in mind that there are true propositions to which no one assents.

6.28

"First we must know what the controversy is about in order to establish what its participants are saying" [Chwedeńczuk 1984, p. 113]. I hold this heuristic principle to be as appropriate as it is simple. For this reason, it is absolutely necessary in analysing the arguments surrounding truth that we separate out the argument about nature from that of criteria — and thus consider them separately.

The question of the criteria of truth can only be decided after deciding the question of the meaning of the words "is true" even if this be merely based on «provisional» intuition.

On the other hand, attempts to explicate "is true" can only be made after deciding that at least certain propositions are true, and others are not.

For here the philosopher has a duty not to lay down but to discover effective methods of establishing when (and perhaps also to what extent) given epistemic procedures justify their objectivity as well as in what circumstances (and perhaps also with what degree of certainty) we can assent to given propositions as true. No doubt, these methods awaiting discovery are applied unconsciously by some and perhaps even consciously by others. The point is nether to seek statistics nor to establish how widespread some taxonomic or research practice is. Indeed, nor is the point to ascertain how such and such understand "truth" and how they in fact determine what is true. (The findings made by lexicographers or «scientio-sociologists» can have some auxiliary significance for semantic analyses.) This is not to say, however, that the choices are arbitrary. To the contrary, they are made under non-trivial constraints.

Thus no one of the contemporary participants in the debate can hold that he is not interested in establishing conditions of the objectivity of cognition — of cognizing things *as they are*. If someone says at the beginning, as the sceptic does, that it is impossible to know things as they are then s/he just places her/himself *hors de combat*.

None of the contemporary participants in the debate can therefore hold that they are not interested in establishing the conditions of truth of true propositions, i.e. of propositions that take account of objective knowledge. Whoever says at the beginning that there are no such propositions leaves the field of epistemology — or at least ceases to be occupied with one of the major epistemic value — but does not thereby supply other solution to the problem of truth.

The sceptic does indeed make a free choice but these choices hardly exhibit any philosophical (non)commitment. Anybody who takes it upon her/himself to take part in the controversy about the problem of truth must find that he has an absolute freedom of choice. The sceptics are not interested in establishing effective methods. However, I cannot bring myself to believe that they are *de facto* interested neither in the truth of any proposition nor in distinguishing objective from non-objective knowledge. They just do not want to proceed beyond intuitive act. And that is all they are allowed to do.

I think there is no basic difference between the way in which the philosopher who is examining the nature of truth proceeds and that in which e.g. chemists proceeded once in examining the nature of acidity. The chemists equally did not use opinion polls to establish the basis of acidity and discover whether something is an acid or not. Moreover, not even linguists do this when they establish the content of the word "acid". In spite of this, the choice of this or that explication of "acidity" is neither a case of someone's axiological hierarchy nor of free choice. Quite simply, certain objects *have* certain properties, which must be described — however we may call them: "acidity" or "falcidity". So here: there *are* true propositions; there *are* certain objective cognitive procedures. They are there independently of the names we choose for them. "For it is an empirical fact" — as Moore might have said. It happens that we assent to true propositions. It happens that we obtain objective knowledge. The theoretical problem consists in how best to formulate explicitly the relevant definitions and characterisations, so that we might when we so wish or others might, understand these things. So that we might in particular understand that meaning of "is true", "where — as Moore says — a true proposition is always true".

6.29

Einübung im Christentum written by Anticlimacus in 1850 contains the following fragment:

> *Denn was ist Wahrheit?* [...] *Die*[*se*] [...] *Frage tat, wie bekannt Pilatus; eine andere Frage ist, ob es ihm eigentlich daran lag diese Frage beantwortet zu bekommen* [Kierkegaard 1850, p. 175].

I have assumed that we philosophers are most certainly interested in the answer, and I hope that the ideas I have just presented were in general satisfactory.

7. ON LINGUISTIC CATEGORIES

7.1

I have the impression that the opulent development of the formalism of the categorial grammar is not accompanied by comparable efforts directed towards the explication of the key concepts of this grammar — with the notion of a semantic category at the forefront. These concepts are usually explained in a textbook fashion, i.e. in a noncommittal, informal and simplified way. However, the matter is both interesting and important enough to merit a careful look at it.

In his essay "On the Problem of Universals", Kazimierz Ajdukiewicz «explains» the conception of a semantic category in ordinary language thus:

(1) Expressions *A* and *B*, taken in sense *a* and *b* respectively, belong to the same semantic category *iff* every sentence Z_A containing expression *A* in sense *a* upon replacement of *A* by *B* taken in sense *b* (the meaning of all other expressions and their interconnections remaining unaltered) is transformed into an expression which is also a sentence, and *vice versa*: every sentence Z_B upon replacement of *B* by *A* (with analogous qualifications) is also transformed into a sentence [Ajdukiewicz 1934d, p. 95].

But in the not much later essay "Syntactic Connexion" we read:

(2) The word or expression *A*, taken in sense *x*, and the word or expression *B*, taken in sense *y*, belong to the same semantic category *iff* there is a sentence (or sentential function) S_A, in which *A* occurs with meaning *x*, and which has the property that if S_A is transformed into S_B upon replacing *A* by *B* (with meaning *y*), then S_B is also a sentence (or sentential function). (It is understood that in this process the other words and the structure of S_A remain the same.) [Ajdukiewicz 1936b, p. 119].

We can find a very similar definition in his *Pragmatic Logic*, published (in Polish) posthumously (cf. [Ajdukiewicz 1965b, p. 19]).

Let us ignore the stylistic differences as well as the addition of alternative words "word" (as an alternative of "expression") and "sentential function" (as an alternative of "sentence"). The essential disparity between (1) and (2) concerns the quantifiers of the *definienses*: in (1), the quantifier is universal while, in (2), existential.

N.B. In some authors' formulations, this disparity is camouflaged in such a way that both the interpretations are admissible. In Witold Marciszewski's "Categorial Grammar", we read about semantic (here: syntactic) category:

(3) Two expressions belong to the same syntactic category *iff* a sentence which contains one of these expressions does not cease to be a sentence if this expression is replaced in it by the other [Marciszewski 1981, p. 43].

According to Alfred Tarski's principle, it should be the case that:

(4) If two expressions are interchangeable in some one context, well-formedness being preserved, then they must be interchangeable in all contexts; hence their interchangeability in at least one context suffices for their being in the same category [Marciszewski 1981, p. 47].

This principle, however, has no application to ordinary language.

I think Ajdukiewicz's alternative explanations will become more clear after assuming that:

(a) we take into account only expressions that have one sense determined in advance;

(b) the expression W contains the expression W^* *iff* the expression W^* is a constituent of the expression W;

(c) everything that is expressed by the terms "replacement" and "transformation" is expressible by the syntactic terms "without" and "identity". Formula (1) can be expressed thus:

(5) Expressions A and B belong to the same semantic category *iff* for every expression Z_A and for every expression Z: if A is a constituent of Z_A, and B is a constituent of Z_B, and Z_A without A is identical with Z_B without B, then Z_A is a sentence *iff* Z_B is a sentence.

Formula (2) can be rendered:

(6) Expressions A and B belong to the same semantic category *iff* there is a certain expression Z_A and a certain expression Z_B such that A is a constituent of Z_A, and B is a constituent of Z_B, and Z_A

without A is identical with Z_B without B, and Z_A is a sentence *iff* Z_B is a sentence.

Now we can easily see that the explanatory value of both formulas depends on the intelligibility of the terms "sentence" and "constituent".

7.2

The first of those terms, i.e. the term "sentence", is explained by Ajdukiewicz in the following manner:

(7) An expression taken in some of its senses is a sentence if — taken in that sense — it is either true or false [Ajdukiewicz 1934d, p. 95].

I define "truth" and "falsity" as follows:

(8) If the sentence S refers to the event Z then S is true (*resp.* false) *iff* Z occurs (*resp.* does not occur).

In his essay "Syntactic Connections between Constituents of Declarative Sentences", Ajdukiewicz defines "sentence" in a different way than indicated in formula (7):

(9) [A sentence is an expression that] denotes a truth-value, [i.e.] truth or falsity [Ajdukiewicz 1960, p. 275].

Here "truth" and "falsity" have senses analogous to the senses of the expressions "occurring event" and "non-occurring event", respectively. Answering the question, what is a sentence and — in consequence — what is a semantic category, is possible only on condition that the semantic term "referring" as well as the ontological terms "event" and "occurrence" are intelligible.

The second of the crucial terms of *definiens* of a "semantic category", i.e. the term "constituent", has not been defined by Ajdukiewicz at all. So it appears that his explanation is not a definition at all, since it includes an intuitive element.

7.3

Let us note at once that formulae (1) and (5), i.e. the formulae with universal quantifiers, do not provide an effective criterion of categorial identity since it is impossible to determine whether every pair of sentences that contains the expressions in question satisfies the conditions or not.

On the other hand, formulae (2) and (6), i.e. the formulae with existential quantifiers, are inadequate because they fail to express some important intuitions. Let us consider the sentence "The cat meows". If we insert in that sentence the expression "the dog barks and the cat" instead of the expression "the cat", we will get the sentence "The dog barks and the cat meows". However, the expressions "the cat" and "the dog barks and the cat" do not belong to the same semantic category. This example can be questioned if we impose such conditions on being a constituent of an expression (or on being an expression in general) that "the dog barks and the cat" fails to be a constituent, *resp.* an expression.

This doubt can be, however, reconstructed by means of the example "Ruthenians and Tartars dispersed". If we insert here the expression "Poles attacked" instead of the expression "Ruthenian", we will get the sentence "Poles attacked and Tartars dispersed". We feel intuitively that the expressions in questions are of quite a different semantic category: "Ruthenians" is a name whereas "Poles attacked" is a sentence. Of course, we can try to save formulae (2) and (6), by claiming that the meaning of "and" is different in these two contexts: it is «conjunctive» (in the first context) and «causal» (in the second).

I am pretty sure that a further counterexample free from such an objection can be constructed.

7.4

A different way of explaining what a semantic category is appeals to the meaning context instead of the sentential context. Marciszewski writes in his essay "Kategoria semantyczna" [Semantic Category] from *Mała encyklopedia logiki* [Little Encyclopaedia of Logic]:

(10) Expressions *X* and *Y* belong to the same semantic category of language *J* *iff* upon replacement of *X* by *Y* (or *vice versa*) in any meaningful expression in language *J*, the formed expression remains meaningful [Marciszewski 1970c, p. 99—100]).

Let us note that the expression "any meaningful expression" in formula (10) admits (especially in its Polish version) of weak as well as of strong interpretation: "an expression" or "every expression".

When is a given expression meaningful? Marciszewski answers this question in his essay "Sensowność" [Meaningfulness] in the following way:

(11) *W* is a meaningful expression of language *J iff* it fulfils one of two conditions:
(a) *W* is a simple expression;
(b) *W* is a complex expression such that all of its simple constituents of it fulfil condition (a), where *W* is syntactically coherent, i.e. constructed according to the syntactic rules of the language *J* [Marciszewski 1970f].

Formula (11) and so formula (10) offers an effective criterion provided that we know: (a) how to decide what is a simple expression; (b) how to decide what is a constituent of a given expression; (c) what kind of syntactic rules are binding on us.

As far as a condition (a) goes, Ajdukiewicz doubts that it is the possible to construct the criterion of simplicity. He writes: "At times, it is uncertain what would be considered as *one* word" [Ajdukiewicz 1936b, p. 121]. However, he allows the possibility of a structural criterion for at least some expressions. He writes: "If a language, e.g. German, admits separable words [...], the criterion for a *single* word cannot be stated structurally" [Ajdukiewicz 1936b, p. 122].

Marciszewski in *Mała encyklopedia logiki* gives us, on the other hand, the following explanations:

(12) [A given expression of language *J* is simple *iff* it is] an expression that belongs to the vocabulary of *J* [Marciszewski 1970a, p. 268].

(13) [A given expression is simple *iff*] no part [*scil.* no constituent] of such an expression is an expression [Marciszewski 1970a, p. 356].

As far as condition (c) is concerned, the knowledge of the syntactic rules can be identified with the knowledge of the quantity and the category of arguments which particular functors involve.

Thus conditions (a), (b) and (c) are fulfilled only if we know the full vocabulary of a given language, the complete list of functors together with the complete structural characteristics, and only if we know how to recognise whether a certain «physical» part of a given expression is a constituent of that expression.

The idea is to avoid recognising i.a. the expression "swallow" as an expression that have the expression "wall" as its constituent, because in the light of formula (13) one would then have to consider the expression "swallow" to be complex.

7.5

As it turns out, all conceptions of semantic category considered above appeal to the notion of a "constituent". Let us assume on trial that:

(14) *C* is a constituent of a meaningful expression *W iff C* is such a part of *W* that (a) *C*, (b) the part of *W* proceeding *C*, and (c) the part of *W* following *C* are meaningful expressions.

This is not a satisfactory definition because it licenses to consider the expression "hat" to be a constituent of the expression "hatred". It would be necessary to amend formula (14) with the rule (R-1) given below. But apart from this, the question arises how to know whether a given expression is a meaningful expression.

Formula (14) cannot be used to make Marciszewski's solution more precise because it would result in the indirect *circulus in definiendo*: the definition of a "meaningful expression" in formula (10) appeals just to the notion of a "constituent".

The fault of *circulus in definiendo* arises also by adding formula (14) to Ajdukiewicz's conception. The reason is that he gives the explication of the notion of "meaningfulness" only for complex expressions, and such an explication assumes the notion of a "constituent".

Also in this case, Ajdukiewicz's explanations are very unsatisfactory. According to them:

(15) [A given complex expression is meaningful *iff* it has] a uniform sense [or a unified meaning] [Ajdukiewicz 1934d, p. 97; 1936b, p. 118].

On the other hand:

(16) A [complex] word pattern [is] a syntactical nonsense [i.e. it] loses unified sense [*iff* it is an] incoherent word pattern [or a loose string of words] [Ajdukiewicz 1934d, p. 98; 1936b, pp. 118, 119].

According to Ajdukiewicz:

(17) [A given complex expression may not be meaningless even if] one would regard it as somewhat unnatural, [or when] it does not make sense [1934d, p. 96; 1965b, p. 19].

I see only one way of giving an account of the uniformity of a certain expression: i.e. by saying that this expression refers to a certain sufficiently isolated object. Thus we are back again to semantic ("reference") and ontological ("sufficiently isolated object") notions.

7.6

Let us assume that we are able to distinguish constituents in a given complex expression. In determining the semantic categories of those constituents we ought to respect the following rule:

(R-1) Exactly one constituent of each complex expression is the main functor (or shortly speaking: the functor) of this expression; the remaining constituents are arguments of this unique functor.

But how to determine which of the constituents of, for example, the expression "the father of the king" is the functor (of category *n/n*)? Is there a criterion for distinguishing descriptive symbols from structural ones or *categoremata* from *syncategoremata*?

Adam Nowaczyk in *Logiczne podstawy języka* [Logical Foundations of Language] answers this question negatively:

(18) The classification of symbols into descriptive and structural ones is arbitrary as far as it goes [Stanosz and Nowaczyk 1976, p. 68].

Ajdukiewicz takes a different position here. First, he claims that:

(19) The semantic category of a single word is defined by its meaning [Ajdukiewicz 1936b, p. 120].

And further:

(20) Any word of a language can be assigned to a certain semantic category by its meaning [...]. Composite [*scil.* complex] expressions may be analysed according to the pattern "functors and their arguments" only if this assumption holds [Ajdukiewicz 1936b, p. 128].

Secondly, according to Ajdukiewicz:

(21) In every significant composite [i.e. meaningful complex] expression it is indicated, in one way or another, which expressions occur as arguments, and to

what expressions, appearing as functors, they belong [Ajdukiewicz 1936b, p. 121].

What is more:

(22) A semantic pattern in which these interrelations are not indicated at all, or only incompletely, is devoid of any coherent meaning [Ajdukiewicz 1936b, p. 122].

Formula (21) does not determine, however, the way, in which it is indicated that a given constituent is a functor (*resp.* an argument) in particular.

We read in Ajdukiewicz about functors:

(23) [Functors are] expressions which combined with other expressions form composite [*scil.* complex] expressions with uniform sense [Ajdukiewicz 1934d, p. 97].

Marciszewski put the matter in a very similar manner

(24) Functors are expressions which serve to form compound [*scil.* complex] expressions out of simple (or less compound) ones [Marciszewski 1981, p. 43].

But the content of the terms "combination" or "formation" is very loose, and they can be used to describe the role of functors as well as their arguments in a complex expression.

This objectionable consequence is avoided by Marciszewski in his essay "Funktor" [Functor]:

(25) A functor is an expression, which is neither a sentence nor a name, and serves to construct either sentences or names, or other (more compound) functors [Marciszewski 1970b].

But formula (25) works only under the condition that we already know how to identify a given constituent (or a given expression) as a sentence or as a name, i.e. how to identify the property of belonging to a basic semantic category.

7.7

As we have seen, the term "sentence" is ultimately defined in semantico-ontological terms.

The term "name" in ordinary language is defined by Ajdukiewicz in such a way:

(26) [The class of names is] that class of expressions each of which belongs to the same semantical category as a noun when used in its normal nominal sense [Ajdukiewicz 1934d, p. 96].

The value of this definition depends of course on the intelligibility of the term "noun", which belongs to the terminology of the traditional grammar. From the logical point of view, this terminology leaves much to be desired.

Ajdukiewicz tries to avoid this troublesome connection by the following reformulation:

(27) These English language expressions which (for some of their meanings) are in the same syntactical [*scil.* syntactic] category as the word "(the) dog" are, in those their meanings, called names in that language [Ajdukiewicz 1965b, p. 19].

However, this definition presupposes, first of all, such a criterion of being an expression that the word "dog" in the context "The ___ barks" not be considered the same expression as the word "dog" in the context "the mark of the ___ is that it barks". Second, this definition again presupposes the problem of categorial identity has been solved.

The last difficulty is avoided in a posterior explanation given by Ajdukiewicz, where he defines "name" directly in the semantico-ontological apparatus:

(28) [Singular names are] those [linguistic expressions] which denote particular objects [Ajdukiewicz 1960, p. 275].

In that context, he also presents a semantico-ontological (not a syntactic, as given in formula (20)) definition of a "functor"):

(29) [Functors are] linguistic expressions which denote functions, [i.e.] relations which to every [...] object of a certain kind assign another object as their correlate in a unique way [Ajdukiewicz 1960, p. 276].

It turns out, then, that a normal (not an ostensive) definition of such key terms of the categorial grammar (for ordinary language) as "(meaningful) expression", "simple expression", "complex expression", "constituent (of a complex expression), "sentence", "name" and "functor" requires an appeal to semantico-ontological terms: "referring", "event", "occurrence", "object sufficiently isolated", "relation", "(ontic) correlate" and "particular object".

In short, such a definition requires «ready-made» semantics and ontology.

7.8

As far as semantics is concerned, the first problem is the distinction of various types of referential function (or denoting) played by expressions that belong to a given semantic category. The point is that expressions of different semantic categories can «refer» to one and the same ontic object (*scil.* correlate). For example, let that object be the event that consists in the fact that Leonardo is painting Mona Lisa. Now the sentence "Leonardo is painting Mona Lisa" as well as the names "the fact that Leonardo is painting Mona Lisa" and "the painting of Mona Lisa by Leonardo" «refer» to that event. By analogy, one can refer to the relation, to which the functor "is painting" «refers» in a certain manner also by means of the name "painting".

I am inclined to assume that the specific function of sentences is to express (resp. to state) beliefs, whereas the specific function of names is signifying objects of any kind. However, it remains a puzzle what the difference is between the fact that a sentence expresses a belief (a certain kind of object after all) and the fact that a name refers to that same belief.

The second difficulty for semantics is the question of the correspondence between the ontological structure of objects and the syntactic structure of expressions of ordinary language that refer to those objects in a certain way. The problem is to decide whether expresssions of the same semantic category that have a different structure but that refer to the same thing (e.g. the sentences "Leonardo is painting Mona Lisa" and "Mona Lisa is painted by Leonardo") can be considered to refer to the same ontic correlate.

This difficulty should be distinguished from the hurdles of the structural analysis of a complex expression, in particular one composed of more than two constituents. The problems here are connected with the phenomena of vagueness and incompleteness of the natural language.

7.9

Ajdukiewicz notices that:

(30) The vagueness associated with everyday speech [...] makes it [...] possible to establish unequivocally [expressions of that language] [...] in a number of ways [Ajdukiewicz 1934d, p. 99].

We can give an account of that vagueness by ascribing more than one admissible syntactic structure to a vague expression. But let us suppose that we deal with an expression that is not vague. Ajdukiewicz writes of such expressions:

(31) Every meaningful and unambiguous expression, consisting of more than one word [...] can be decomposed without any residue into its components in exactly one way, so that one of those components [namely the functor or the operator] relates to others [namely arguments] and connects them into a meaning[ful] whole [Ajdukiewicz 1960, p. 270].

The fulfilment of the condition of uniqueness of admissible structural analysis in the case of expressions of ordinary language requires complementing rule (R-1) with several constraining rules: (R-2)—(R-4).

(R-2) In a complex expresssion that belongs to a basic semantic category, at least one argument of a functor belongs to a basic semantic category.

According to rule (R-2), in the expression "He really knows" one may isolate neither the word "knows" as the functor, and the sequence "he really" as its argument, nor the word "really" as the (main) functor, and the words "he" and "knows" as its arguments. This is because neither the sequence "he really" nor the word "knows" is a name or sentence. Only one interpretation remains:

HE REALLY KNOWS
n *s/n//s/n* *s/n*
s/n

(R-3) At least one argument of a complex functor belongs to the same sub-category as that functor.

Let us consider the sentence "Ladislaus is the father of Casimir". It could be interpreted (i.a.) as:

LADISLAUS IS THE FATHER OF CASIMIR
n *s/nn* *n*

or as:

LADISLAUS IS THE FATHER OF CASIMIR
n *s/nn* *n*

The first interpretation is excluded by rule (R-3) because, in the putative complex functor "is the father of" (*s/nn*), the only argument of the inner functor "is", i.e. the expression "the father of", does not belong to the sub-category *s/nn*. However, that interpretation would be admissible if we recognised the expression "is the father of" as a simple expression.

Similarly, the sentence "Barbara thinks of marriage" admits, more than one interpretation even if (R-1), (R-2), and (R-3) are applied:

BARBARA THINKS OF MARRIAGE
n *s/n* *s/n//s/n*
s/n

and:

BARBARA THINKS OF MARIAGE
n *s/nn* *n*

if we do not decompose the expression "of marriage" in the first case, and the expression "thinks of" in the second one.

The first interpretation could be eliminated, if we assumed *ad hoc* the meta-rule:

(M-R) From among two interpretations admissible on the ground of rules (R-1)—(R-3), choose that interpretation in which the functors are less complex.

(R-4) All arguments of a given functor belong to one and the same semantic category.

Rule (R-4) makes it possible to eliminate the ambiguity of, for instance, intensional sentences of the type "*A* believes that *p*". If rule (R-4) did not hold, they would have two admissible interpretations:

A BELIEVES THAT *p*
n *s/ns* *s*

and

A BELIEVES THAT *p*
n *s/nn* *n/s* *s*
n

Rule (R-4) demands the rejection of the first interpretation.

7.10

Aside from the ambiguity, the second source of difficulties connected with the structural analysis of complex expressions of ordinary language is their incompleteness. The problem is how to distinguish incomplete expressions from nonsenses. Let us define "incomplete expressions" in the following way:

(32) The expression *W* is incomplete *iff W* lacks a constituent of a proper semantic category.

However, according to such a definition one could recognise any sequence of expressions as incomplete, including the sequence "if or, then" because one could «complete» that sequence to an expression with "uniform sense", e.g. to the expression "If an imprudent or abject man sets fire to a cover, then the forest will burn down".

The possibility of such a «disenthymematisation» will not be excluded even if we assume that:

(33) A sequence of expressions is nonsense *iff* it contains more constituents than are «needed» or if at least one of them belongs to an «inappropriate» semantic category.

I suspect that this problem cannot be solved without appeal to pragmatic criteria.

7.11

Thus, the conclusion of the above remarks is this.

In order to construct a categorial grammar for ordinary language (where it is to be more than an arbitrary enumeration of the vocabulary appended with categorial indexes, including sufficiently characterized functors), one would have to either (a) enrich the list of rules of the categorial interpretation (and meta-rules) beyond (R-1)—(R-2) and (MR), or (b) appeal to a «ready-made» semantic (and probably also pragmatic) theory as well as to ontological theories of competent speakers.

It can hardly be taken for granted that the first condition is attainable. In view of the second one, the putative universality of categorial grammar would be derivative of the universality of semantic and pragmatic relations as well as of the universality of the ontological «picture» of the world.

8. ON QUESTIONS

In 1964, during a meeting of the Cracow Division of the Polish Philosophical Society, Roman Ingarden delivered a paper under the dramatic title: "What do we not know about values?". If the measure of the drama of the situation were the length of the list of problems needed to be addressed in the answer, it would be equally dramatic to ask today what we do not know about questions.

Erotetics has a long tradition. In Poland, it reaches as far back as Kazimierz Twardowski's time. We have at our disposal formal interpretations concerning some of the main erotetical intuitions. However, we are still unable to give a satisfying answer to such questions as: What is a question, and when is it well-posed? What is an answer to a given question, and when is it proper? What is expressed by a question, and what does a question refer to? What is assumed by a question? Examples can be multiplied...

Let us try to deal with over some difficulties of «proto-theoretical» erotetics.

8.1

Sometimes, the question "What is a sentence?" is answered: it is an expression that is either true or false. I am not satisfied with such an answer. For I define "truthfulness" (reductively) as follows:

If a sentence S states that p then if the sentence S is true then p.

Searching for an answer to the question mentioned above, people appeal either to a semantic or to a pragmatic function of sentences.

Someone who defines "sentence" by an appeal to its semantic function usually claims that a given sentence is identical with an expression that states a certain state of affairs. A definition of this kind is instructive, when we know from another independent source what the function of stating consists in and what a state of affairs is. Let us leave aside the ontological troubles concerning the separation of states of affairs from other ontic categories. As to the function of stating, let us note that it is not an «ordinary» correlation. With a certain state of affairs one can correlate not only a sentence but also a name. For example, with the fact that the sun is shining, one can correlate the name "the fact that the sun is shining" as well as the sentence "The sun is shining". It is claimed that while the sentence "The sun is shining" states this fact, the name "the fact that the sun is shining" designates it. But how can we distinguish stating from designating PRIOR TO distinguishing sentences from names?

Someone who defines "sentence" by an appeal to its pragmatic function, usually claims that a sentence is identical to an expression that expresses a certain belief. Again, this definition is instructive provided that we know (independently) what the function of expressing consists in and what a belief is. Let us leave aside the psychological problems how to distinguish beliefs from other experiences. Certainly, names do not express beliefs. But how are we to distinguish expressing from stating? Let us try the following. The sentence "The sun is shining" states that the sun is shining but it expresses the fact that a person uttering this sentence — let us call this person "the sender" — is believed that the sun is shining. Clearly, however this sentence does not state that our sender believes that the sun is shining. This last state of affairs is stated by the sentence "The sender of the sentence "The sun is shining" believes that the sun is shining". Similarly, this last sentence does not express the fact that the sender believes that the sun is shining; it expresses the fact that the sender believes that s/he is believed that the sun is shining.

Thus, could a sentence really be defined in terms of its pragmatic function?

8.2

Let us repeat an analogous argument with reference to questions. Let us agree that the word "question" is a short for the expression "interrogative sentence".

Because finding a property of questions analogous to the truth-value of sentences is quite a problem, we shall appeal either to a semantic or to a pragmatic function of questions.

Thus, we can say that a question is an expression, which concerns a SPECIFIC state of affairs. Following Ingarden [1925, pp. 327ff], let us call it a "problem".

«Concerning-a-problem-by-a-question», like «stating-a-state-of-affairs-by-a-sentence», is a correlation but it is not an «ordinary» correlation. Let us consider the problem which the question "What is shining?" concerns. The sentential function "*x* is shining" also concerns this problem, i.e. the problem WHAT is shining.

In this situation, one can try to define "question" by an appeal to its pragmatic function. But what is expressed by a question? Let us call what is expressed by a question — following Ingarden again — an "interrogation". However, what is an interrogation?[1]

Is interrogation a lack of belief? A lack of belief is usually expressed by a sentence of the type "I am not convinced of that-and-that".

Is interrogation a lack of knowledge? Again, a lack of knowledge is commonly expressed by a sentence of the type "I do not know that-and-that"; in our case — "I do not know what is shining".

In any case, it would not be an ordinary ignorance, for it is not due to the lack of knowledge-that. In our case, what is relevant is not an ignorance THAT the sun is shining, but an ignorance about WHAT is shining. Moreover, we do not have a general formula proper to this second kind of ignorance.

Problems with ignorance-that are well known. It appears that we are able to formulate sufficient conditions of such an ignorance:

If not-p or O is not justified in believing that p, then O does not know that p.

But we are not able to formulate its necessary condition. The same applies to ignorance in case of questions.

[1] According to Kazimierz Ajdukiewicz the state of questioning — i.e. a question in the psychological sense — is "a mental tension, similar to thirst; it is a state in which [...] [a] person strives to develop a conviction that may be expressed by a proper answer to that interrogative sentence; [...] to acquire a certain information; [...] is A SPECIAL form of interest" [Ajdukiewicz 1965, pp. 91—92]. The obscurity of Ajdukiewicz's formulations is not accidental: we are facing real problems here.

However, the crux of the matter consists, first of all, in the fact that this ignorance is at most one of the necessary elements of interrogation. When we ask (let us ignore didactic questions), there is something that we do not know and, moreover, we WANT to know it. When we want to come to know what is shining, we ask: "What is shining?". Thus, this question expresses that we do not know what is shining and, at the same time, that we want to know what is shining. The fact that we want to know what is shining can be stated by the sentence "I want to come to know what is shining"; but this sentence expresses something else: it expresses my belief that I want to know what is shining.

In turn out that one can distinguish a situation in which our desire to know something is stated (by means of a suitable sentence) from a situation in which this desire is expressed (by means of a suitable question).

Thus, it looks as if a question could also be defined only by means of its pragmatic function.[2] However, the matter is not quite so clear because the volitional component of interrogation can be expressed not only by a suitable question but also by a suitable order. The fact that I want to know what is shining can be expressed by the demand "Tell me what is shining!" as well as by the question "What is shining?".[3]

8.3

Prima facie, the satisfactory answer to the question "What is a *datum questionis*?" is: it is a sentential function created from a given sentence by substituting for its question-marker a variable of the same semantic category.[4]

Indeed, the *datum questionis* of the question "What is shining?" is a sentential function "*x* is shining" which results from this question by

---

[2] This is what Twardowski has pointed out by saying that a question is a CALL for an utterance [Twardowski 1901, p. 158].

[3] After all, Leon Koj has showed that no attempt at paraphrasing questions by means of sentences — whether sentences "grasping the interrogator's knowledge", or sentences "expressing his doubts or ignorance", or sentences "revealing his desire to acquire information" — is satisfactory [Koj 1971, p. 113].

[4] Sometimes the term "CONTEXTURE" is used instead of "*datum questionis*" [Poletyło 1970, pp. 232—234]. But much more often, this term names the part of question remaining after the removal of the question-marker.

substituting the variable "*x*" of nominal category for the question-marker "what".

Unfortunately, such an approach can hardly be considered satisfactory even for simple questions.

Sometimes, it is hard to guess what the question-marker is. Consider a question like "Which Pole fought for the liberty of America?". If it is "which Poles" then the *datum questionis* is "*x* fought for the liberty of America"; if it is "which" then the *datum questonis* is "*x* Pole fought for the liberty of America". We will ignore this problem. After all, the matter can be settled conventionally.

What is a problem is that having at our disposal only a *datum questionis*, we can not reconstruct the initial question without additional assumptions. After all, the *datum questionis* of the question "What is it?" has the same form as the *datum questionis* of the question "Where is it?", viz. the form of the function "It is *x*".

Perhaps, the *datum questionis* of the question "What is it?" should have the form of a function with a semantic category of its variable explicitly marked, i.e. "It is *x*(-*n*)". But how can we know that the question-marker "what" determines just this semantic category of a variable? To reach certainty in this matter, it would be necessary to construct a lexicon correlating a definite semantic category of each variable with all question-markers of a given language.

8.4

Let us agree that the *datum questionis* is a sentential function that correspond to a problem (in Ingarden's sense), which a given question concerns. We should start with the analysis of the problem so understood by establishing which *datum questionis* a given question has.[5]

Thus:

(a) Questions "WHAT has the property *P*?" concern problems that consist in the fact that *x*(-*n*) has the property *P*. Let us call such problems "SUBSTANTIAL problems".

[5] Marek Tokarz goes further in this direction; he notices that "one does not have to construct a special theory of questions because one can study sets of sentences of a definite structure instead of dealing with questions; then the logic of questions would become a logic of answers" [Tokarz 1984, p. 167].

(b) Questions "WHEN does *a* have the property *P*?" concern problems that consist in the fact that *a* has the property *P* in the period *t*. Let us call such problems "TEMPORAL problems".

(c) Questions "WHY does *a* have the property *P*?" concern problems that consist in the fact that *a* has the property *P* because *q*. Let us call such problems "CAUSAL problems".

(d) Questions "WHAT does *a* have the property *P* FOR?" concern problems that consist in the fact that *a* has the property *P* in order β. Let us call such problems "TELIC problems".

The variable "β" in the *datum questionis* that concerns a telic problem is an infinitive and it has the form "have the property *Q*", where *a* is the possessor of this property. In the case of the sentential function "*a* lies in order β", the infinitive "to survive" can be substituted for "β", where this infinitive is an «equivalent» of "that *a* survives".

What is the semantic category of this infinitive? If following Ajdukiewicz, we assume only two basic categories (*scil.* sentences and names) then this infinitive would probably be a functor. But which functor? In the context "To survive (in such-and-such conditions) is a great trick", the infinitive "to survive" is a name. In the context "*a* has to survive", the infinitive "to survive" is a functor *s/n//s/n*. If one assumes that the infinitive "to survive" is such a functor in the context "*a* lies in order to survive", then "in order [to]" is a functor *s///ss/n//s/n*.

Ingarden divides questions into existential and matter-of-fact questions.[6] A real difficulty arises when one wants to indicate what an EXISTENTIAL problem is, i.e. the problem that the decision-question "*p*?" concerns. We seem to expect "Yes" or "No" answers to such questions. However, such answers seem to be shorthand for the sentences "*p*" and "It is not the case that *p*", respectively. For the sake of symmetry, let us agree to treat the sentence "*p*" as a shorthand for the sentence "It is the case that *p*". Moreover, the expressions "it is the case" and "it is not the case" can be treated as expressions of the semantic category *s/s*. Thus, one could argue that the *datum questionis* could have the form "*x*(-*s/s*) *p*". But what is a question-marker in a decision-question, or — respectively — what is unknown in an existential prob-

---

[6] Ajdukiewicz propagated Twardowski's terms "decision-questions" and "complementation-questions", respectively.

lem? Ingarden would have certainly answered: the so-called existential moment. But how can we express it in «our» terms?

8.5

We read in Ajdukiewicz:

> By a *positive assumption of a question* we mean the statement that at least one proper answer to that question is true, which is equivalent to the disjunction of all proper answers to that question [Ajdukiewicz 1965b, p. 88].

Before, we come to know that:

> The sentences obtained from the *datum questionis* of an interrogative sentence by the substitution for the unknown of an appropriate constant are termed *proper answers* to that interrogative [...] [sentence]. The question-markers (with possible modifiers) [...] usually single out those values of that unknown which, when substituted for the unknown, turn the *datum questionis* into a true or a false statement. [...] It singles them out in the sense that only [...] [these values], when substituted in the *datum questionis* for the unknown of the question, turn the *datum questionis* into a proper answer to that question. [...] The set of the values of the unknown of a question which are singled out by the question-markers are termed the *range of the unknown of the question* [Ajdukiewicz 1965b, pp. 86—87].

Thus, we may say that:

(*) a proper answer to a given question is an answer, which is a result of the substitution of one of the distinguished values for the variable in the *datum questionis*.

But what is a distinguished value?

To say (as Ajdukiewicz did in the quote above) that a distinguished value is such a value that if we substitute it for the variable in the *datum questionis*, we shall get a proper answer, does not say much because it only reverses the dependence (*). A distinguished value has the same semantic category as the variable to whose range this value belongs. However, such a characterisation is not sufficient either. The expression "it is a good thing that" is of the *s/s* category but we would not recognise the sentence "It is a good thing that *p*" as a proper answer to the decision-question "*p*?".

Again, nothing remains for us to do but to make a lexicon that correlate not only a definite semantic category with a variable but also ranges of variability (i.e. sets of distinguished values) with all the question-markers of a given language. This would be consistent with the directive that only schemata that contains variables with DEFINITE ranges of variability be treated as sentential functions. This in turn is tantamount to Twardowski's directive, according to which questions ought to be UNIVOCAL [Twardowski 1901].[7]

Let us complete the notion of a positive assumption of a question with the notion of a negative assumption of a question. We read in Ajdukiewicz:

> By a *negative assumption of a question* we mean the statement that at least one proper answer to that question is false, which is equivalent to the disjunction of negations of all proper answers [Ajdukiewicz 1965b, p. 88].

And now, let us consider a question of the type "WHAT has the property *P*?" concerning a substantial problem, i.e., having the *datum questions* "*x*(-*n*) has the property *P*". Let *U* be the range of variability of the unknown of this question. Thus, we can say that the sentence:

$(Ex)\ K\ (x \in U)\ (Px)$

is the positive assumption of this question, whereas the sentence:

$(Ex)\ K\ (x \in U)\ N\ (Px)$

is its negative assumption. One can say, in general, that the assumption of our question is the following conjunction:

$K\ (Ex)\ K\ (x \in U)\ (Px)\ (Ex)\ K\ (x \in U)\ N\ (Px).$

At first sight, everything seems all right. When we ask "Who discovered America?" (an example given by Ajdukiewicz) we seem to assume that someone discovered America and that not all people discovered it.

However, is this always the case with substantial questions? If we pose the "Who wants to go to San Remo?" to a group of persons, which constitutes, in this case, the range of variability of this question, do we really assume that somebody but not everybody wants to do it? Do we consider the answers "All of us" and "Nobody" unsuitable? It seems that considering them as unsuitable would only extend properties of didactic questions over all of the substantial questions.

---

[7] This is just what Tadeusz Kubiński did who interpreted question-markers (in complementation questions) as special EROTETIC quantifiers [Kubiński 1971].

In the case of a decision-question concerning an existential problem, i.e. having the *datum questionis* "x(-s/s)p", the range of variability of its unknown contains two values: "it is true that" and "it is not true that".[8]

In any case, assuming the principle of bivalence, the assumption of such a question would be a conjunction of the form:

$K\ (Ex)\ K\ (x \in U)\ (xp)\ (Ex)\ K\ (x \in U)\ N\ (xp).$

The first conjunct will be satisfied by one of the values while the second one will be satisfied by the other. And the assumption of the existential question is true *ex definitione* because precisely one of the two possibilities occurs: either it is the case that *p*, or it is not the case that *p*.

However, let us consider a question of the type "WHY does *a* have the property *P*?", concerning a causal problem, i.e. having the *datum questionis* "*a* has the property *P*, because *q*". Here, causes are distinguished values. But is this true for all causes? If we accept effectualism, according to which every state of affairs has a certain effect, then are all the states of affairs relevant here? Is the sentence "The Sun is shining because the Moon is a slice of cheese" a proper answer to the questions "Why is the Sun shining?"? And are the real states of affairs (e.g. the fact that Poland joined NATO) the only relevant ones? Or maybe it is only the «possible» causes of the fact that the Sun is shining (and in general: the fact that *a* has the property *P*) that should be taken into account? In any case, we would have to compile the full repertoire of such possible causes for each state of affairs.

The case of telic questions present us, *mutatis mutandis*, with the same difficulty.

8.6

We read in Ajdukiewicz:

---

[8] However, are these just two truth-values, as Jerzy Kopania [1987, p. 297] writes? We ignore here the difficulty connected with the fact that sometimes "*p*" is a hidden conjunction, as in the case of the question "Has George stopped stealing yet?", which, in fact, stands for two questions: "Did George formerly steal?" and "Does George not steal anymore?" [Tokarz 1984, p. 165].

> A question whose positive or negative assumption is not true is termed *improperly posed* [Ajdukiewicz 1965, p. 88].[9]

Thus, we want to say that a properly posed question is a question that has a true positive assumption and a true negative assumption, i.e. in short: that has a true assumption.[10]

However, let us now suppose that, for instance, a substantial question has no assumption: neither a positive nor a negative one (in the senses explained above). Suppose also that a definite range of variability is correlated with their question-markers. Due to this fact, such a question would be well-posed. One would expect that a proper answer to such a question states precisely one of the following possibilities:

(i) that no value from the range of variability of its unknown satisfies the *datum questionis*;

(ii) that some, but not all, values from this range satisfy the *datum questionis*;

(iii) that all values from this range satisfy the *datum questionis*. .

An improper answer to a substantial question would state precisely one of the three possibilities:

(i′) that no value from a certain set different from the range of variability of its unknown satisfies the *datum questionis*;

(ii′) that some, but not all, values from this set do not satisfy the *datum questionis*;

(iii′) that all the values from this set satisfy the *datum questionis*.

But is it the case that ONLY such an answer is improper? An answer that did not «revert» to the constant part of the *datum questionis* of a given question would also seem to be improper. Thus, the answer "It is thawing" would also be an improper answer to the question "What is shining?". And perhaps it is not an answer to this question at all?

For what is an answer?

> A response to a question [...] is [...] [a statement that] to some extent complies with the intention of the person who has asked the question [Ajdukiewicz 1965b, p. 89].

[9] Sometimes, in this context, a proper answer is called "a possible answer" [Kopania 1987, pp. 296—310]. A possible answer should be distinguished from an accurate answer [Twardowski 1901], which contains "nothing less than" is expected by an interrogator.

[10] Twardowski [1901] called such questions "logical questions", i.e. questions that "did not venture to contain wrong data".

Let us return to well-posed questions. According to causalism, each causal question is well posed. On this ground, the answers of type (i) are not acceptable; moreover, according to monocausalism (*scil.* the view that everything has precisely one cause), there are no acceptable answers of type (iii). But they are not acceptable because they are simply false, not because they are improper.

Answers that do not «revert» to the constant part of the *datum quaestionis* of a causal question are improper as was the case with such answers to substantial questions. For instance, the answer "It is thawing" is an improper answer to the question "Why is the sun shining?".

In general, one can say that every expression that is non-grammatical (or not well-formed) is also an improper answer (to any question).

* * *

As we have seen, our intuitions concerning questions of everyday linguistic practice are unclear and vague. If these intuitions were to constitute an intuitive horizon for an adequate logic of questions, they should first be thoroughly explicated and modified.

9. ON SEMIOTIC FUNCTIONS OF CONDITIONALS

9.1

At the beginning of this century it was generally held — e.g. by Władysław Biegański [1903, p. 184; 1912, p. 281] — that a conditional sentence states or, in other words (!), expresses the succession and contiguity (in time and in space) of those states of affairs, to which the antecedent and the consequent of the conditional refer. However, the conditional states (expresses) nothing about these states of affairs taken separately. The introduction by logicians of the concept of material implication has given rise to the question of its relation to the conditionals of natural language. One should note that the question is held to have appeared as long ago as Ancient times [Łukasiewicz 1934, p. 203—204]. I suppose that amongst Polish logicians the first to draw attention to the so-called paradox of material implication was Tadeusz Kotarbiński [1929, p. 168].

Almost at once, two contrary views on the issue became clear. According to one, whose most famous proponent was Roman Ingarden [1936], material implication is, from the semantic point of view, so different from conditional sentences that the latter can in no way be reduced to the former. According to the second, most fully developed by Kazimierz Ajdukiewicz [1936a], material implication and the conditionals of natural language have the same truth-conditions; the paradox of material implication can be avoided once the distinction between "stating" and "expressing" is introduced. This distinction was all the easier to accept due to the previous progress in the analysis of the expressive function of utterances carried out by Maria Ossowska [1928; 1931].

9.2

Because of the interruption of normal academic life in Poland — first as a result of the destructive war, then the cultural barbarism of the occupying powers, and later the distorted exigencies of post-war politics — scholars had to retrace their steps a number of times to the origins of the dispute. First of all, the authors themselves made, at a greater length than previously, reference to the solutions of their pre-war writings: I mean Ingarden [1949] and Ajdukiewicz [1956a]. However, they did not develop their positions any further. Soon, they were joined by two new reserchers: Zdzisław Czerwiński [1958] and, slightly later, Eugeniusz Grodziński [1969]. (I am setting aside the purely expository writing of Ludwik Borkowski [1964] and Janina Kotarbińska [1964].)

Grodziński attempted (unsuccessfully, I believe) to show that a sentence of the form "If *p* then *q*" (where "*p*" and "*q*" are sentences) is a (metalinguistic) sentence about the sentences "*p*" and "*q*". To be more exact, it is a sentence that states that some relation holds between the logical values of "*p*" and "*q*" [Grodziński 1969, p. 64], and moreover, that if it is a true sentence, it expresses a valid (deductively infallible) inference [Grodziński 1969, p. 60].

On the other hand, Czerwiński sought to cast doubt, to my mind correctly, on Ajdukiewicz's analysis of the semantic function of stating in its connection with conditional sentences. He showed that:

(1) the concept of "stating" as expounded by Ajdukiewicz is unclear [Czerwiński 1959, p. 265];

(2) given the limits of the solution put forward by Ajdukiewicz, one can assume that if two sentences state the same thing then they are logically equivalent [Czerwiński 1959, p. 265];

(3) what natural language conditionals states is not the same as what sentences exemplifying material implication state: there are false conditionals in natural language that become true if the "if ... then" functor is interpreted as material conditional [Czerwiński 1959, pp. 265—266]; an example would be "If Copernicus had a son then he was not a father" — as would be every sentence, whose antecedent logically excludes its consequent [Czerwiński 1959, pp. 266—267];

(4) a natural language conditional is true if there exists a true formal implication, from which the conditional can be obtained by substitution [Czerwiński 1959, p. 269].

9.3

After the death of Ajdukiewicz, Barbara Stanosz became the main proponent of his approach to solving the paradox of material implication [Stanosz and Nowaczyk 1976; 1985]. Jerzy Kmita made progress in the analysis of the expressive functions of utterances [Kmita 1966], as did Leszek Nowak [1970]. The task of clarifying the notion of "connection in content" — a key notion of the alternative approach — fell to Zdzisław Kraszewski [1972].

A particularly interesting standpoint on the issue was taken by Jerzy Pelc [1985].

First of all, he drew up conditions of linguistic admissibility of conditionals, i.e. combined conditions of their truth and proper use, in a pragmatically based analysis. According to Pelc, language-users regard a conditional as admissible if:

(1) they believe that there is a connection between the content of the antecedent and that of the consequent;

(2) they do not believe that the component sentences are not true;

(3) they do not believe (are not convinced) that they are true;

(4) they believe that there is a conditional connection (and they «look» for it) between the content of the antecedent and that of the consequent.

Second, as can be seen above, Pelc not only includes amongst his conditions the language-user's belief that a consequence-relation holds between the antecedent and the consequent, but also her/his belief that there is a factual connection between the states of affairs indicated by the antecedent and those indicated by the consequent. The latter belief would seem to be the basis for the former.

9.4

The views of Andrzej Bogusławski must be added, for the moment, to the anti-Ajdukiewicz approach [Bogusławski 1986]. They are in a sense the complement of Czerwiński's position. Bogusławski called into question the second crucial element in Ajdukiewicz's solution: i.e. the conclusions of his analysis of the pragmatic function of «expressing» for conditionals. He tried to demonstrate that:

(1) Ajdukiewicz's notion of "expressing" is unclear;

(2) "expressing" can only be distinguished from "stating" if: (a) the counterdomain of the relation of expressing (and, respectively, of stating) is homogeneous; (b) "expressing" and "stating" have non-relative content;

(3) one has to introduce the concept of "saying" (direct and indirect) in order to explain why the following are held to be deviant: (a) uttering the negation of the sentence "I do not know if p, and I do not know if q" together with a sentence of the form "If p then q"; (b) uttering a conditional of the form "If p then q" in which "p" is a sentence of the form "I know that r";

(4) every (basic) conditional of the form "If p then q" says: (a) directly, that there is a connection between the fact that p and the fact that q; (b) indirectly, that the speaker knows neither whether p nor whether q;

(5) certain (basic) conditionals of the form "If p then q" state directly what is stated by the ideal (i.e. strict) implication: "$p \Rightarrow q$", i.e. that it is impossible that p and not q;

(6) no (basic) conditionals of the form "If p then q": (a) make precisely the same statement as is made by the material implication "$\rightarrow$"; (b) says either directly or indirectly that the speaker knows that p;

(7) the consequence of a correct utterance of a (basic) conditional is that the speaker is willing to infer "q" from "p".

I think that Bogusławski applies conditions which are too rigorous for the separateness of "expressing" and "stating". If it were a condition of the separate extension of two names that no thing belonging to the extension of one should have a property common to something in the extension of the other, then no two names would be separate. In order to distinguish the concept of expressing from that of stating it is enough to show that these relations have different fields. I would say, moreover, that this distinction suffices to clarify the linguistic phenomena referred to by Bogusławski in (3) above. Going further, it seems to me that all difficulties of this kind are cleared up by making use of the expressing-stating distinction. Contrary to Bogusławski, I think that the existence of a connection between the fact that p and the fact that q is *stated* by a conditional of the form "If p then q". On the other hand, the fact that the speaker knows neither whether p nor whether q, is *expressed* by such a conditional. Moreover, a conditional expresses, amongst other things, the fact that the speaker knows that there exists a connection between the fact that *p* and the fact that *q*. Thus, it also expresses his readiness to infer "*q*" from "*p*".

9.5

Now I should like to put forward a solution, which keeps the framework of Ajdukiewicz's approach but it also takes into account Czerwiński's corrections and certain of Bogusławski's conclusions. Hopefully, it will equally avoid the difficulties to which both these scholars drew attention.

(1) The principle of rationality.[1] If *X* is rational then if *X* rejects "*p*", one must not accept that *p*.

N.B. Bearing in mind Alfred Tarski's reservations [Tarski 1933, pp. 159—160], I nevertheless often write for convenience "*X*", "*Y*", etc. as names of individuals *X*, *Y*, etc. and "*p*", "*q*", etc. as names of sentences *p*, *q*, etc.

(2) If one may not accept that *p* then "*p*" is illegitimate or unacceptable.

Indicative, interrogative and imperative sentences may all be illegitimate (unacceptable). "Illegitimate indicative sentence" has the same meaning as "false (incorrectly uttered) sentence"; "illegitimate interrogative sentence" has the same meaning as "unjustified (wrongly put) sentence"; and "illegitimate imperative sentence" has the same meaning as "inapplicable (wrongly delivered) sentence".

(3) If "*p*" is legitimate then if "*p*" states that *q* then *q*.

As has often been emphasised (most incisively perhaps by Pelc; cf. for instance the recent [Pelc 1982, p. 227]), if one is to be exact, one should not say that "*p*" states that *q*, but rather that *X*'s utterance that *p* states that *q*. Similarly sentences asserting that "*p*" announces that, presupposes that, expresses *q*, should be seen as abbreviative.

---

[1] My principle of rationality is similar to H.P. Grice's principle of quality [Grice 1975, p. 46].

(4) If "*p*" states that *q* and "*q*" is identical with "*p*" then "*p*" announces that *q*.

(5) Every "*p*" announces that *p*.

(6) If "*p*" states that *q* and "*q*" is different from "*p*" then "*p*" presupposes that *q*.

(7) If "*p*" is correct, then if "*p*" expresses that *q* then *q*.

(8) If "*p*" states that *q* then "*p*" expresses the belief that *q*.

(9) If "*p*" states that *q* or "*p*" expresses that *q* then "*p*" informs that *q*.

As can be seen, I am endorsing here the distinction of the semantic from the pragmatic function. I think that as long as epistemological dualism remains insurmountable, the opposition of "stated" and "expressed" is justified, where what is expressed is intersubjectively impervious.

9.6

(10) The principle of intelligibility. If *X* understands "*p*" then *X* knows the content of "*p*", i.e. knows the set of states of affairs about which "*p*" informs.

(11) There are various types of conditionals that are mutually irreducible.

The main ones are as follows:

(a) NATURAL IMPLICATIONS;

(I) general implications: "Whenever *p*, *q*", where "*p*" and "*q*" stand for open sentences;

(II) individual implications: "(Even) if *p* then (it is not the case that) *q*", where "*p*" and "*q*" stand for closed sentences;

(A) differentiating implications of the form "If *p* then *q*": (i) factual implications: "Since *p*, *q*", where "*p*" and "*q*" may take any tense; (ii) potential implications: "Supposing that *p*, *q*", where "*p*" and "*q*" may be set in any tense; (iii) counterfactual implications: "If it had been the case that *p* then it would have been the case that *q*", where "*p*" and "*q*" are set in a past tense;

(B) qualifying+ implications (they may be factual, potential or counterfactual) of the form "Even if p, q";

(C) qualifying– implications (factual, potential or counterfactual) of the form "Even if p, it is not the case that q";

(b) LOGICAL IMPLICATIONS;
(I) material (loose) implications: "$p \rightarrow q$";
(II) ideal (strict) implications: "$p \Rightarrow q$".

I realise that, on the one hand, one type of conditional may be uttered by means of various types of conjunctions, and on the other, one conjunction may occur in different types of conditionals. For simplicity I only use certain particular conjunctions in order to mark certain particular conditionals. I am also aware that conditional sentences — certainly those that exhibit factual and potential implication — do not have to be indicative sentences but may also be interrogatives or imperatives.

Below, I limit myself to an analysis of indicative conditionals.

9.7

(12) A conditional of the form "Supposing that p, q" presupposes, amongst other things, that p is connected with q.

If it is obvious that there is no such connection then the conditional in question, to use Ingarden's phrase, is an "arrant nonsense" [Ingarden 1936, p. 266].

(13) A conditional that exemplifies potential implication presupposes a certain connection. The most frequent types of connections are:

(a) symptomatic connection; for example:

If I have a fever then I am ill;

(b) coexistential connection; for example:

If my wife was at Duszniki at that time then she definitely listened to Chopin;

(c) generic connection; for example:

If Copernicus was a Pole, he was a Slav;

(d) significatory connection: for example:

If the light is red, the road is closed;

(e) motivational connection; for example:

If the priest goes there, he will become transfixed with joy;

(f) causal connection; for example:

If you eat hemlock, you will soon die;

(g) necessary connection; for example:

If an area of low pressure occurs in the Mazovia region, it will rain in Warsaw.

In the example of (a), it is presupposed that a fever *requires* (is the physical consequence of) illness. Sentences of the form "If I am not mistaken then *q*" are of a similar type. N.B. Only in the case of a sentence of the form "If *p* then *q*" that states a symptomatic connection, would I be inclined to say that the fact that *q* is a necessary condition for the fact that *p*.

In the example of (b), it is presupposed that my wife's stay in Duszniki WAS ACCOMPANIED BY her listening to (the music of) Chopin.

In the example of (c), it is presupposed that (Copernicus') being a Pole CONSTITUES (his) also being a Slav.

In the example of (d), it is presupposed that the red light SIGNIFIES a closed road.

In the example of (e), it is presupposed that the priest's going there GIVES RISE TO (is the psychological cause of) his being transfixed with joy. Under this heading come all conditional commands, prohibitions, promises and threats.

In the example of (f), it is presupposed that the eating of hemlock CAUSES (physically) a quick death. N.B. Only in the case of a conditional which presupposes a causal connection, would I be inclined to say that the state of affairs indicated by the antecedent is a sufficient condition of the state of affairs indicated by the consequent.

In the example of (g), it is assumed that the occurrence of low pressure in the Mazovia region MAKES IT POSSIBLE THAT it rains in Warsaw. N.B. Only in such a type of conditional, would I say of the state of affairs referred to by the antecedent that it is a necessary condition of the state of affairs obtaining that is referred to by the consequent.

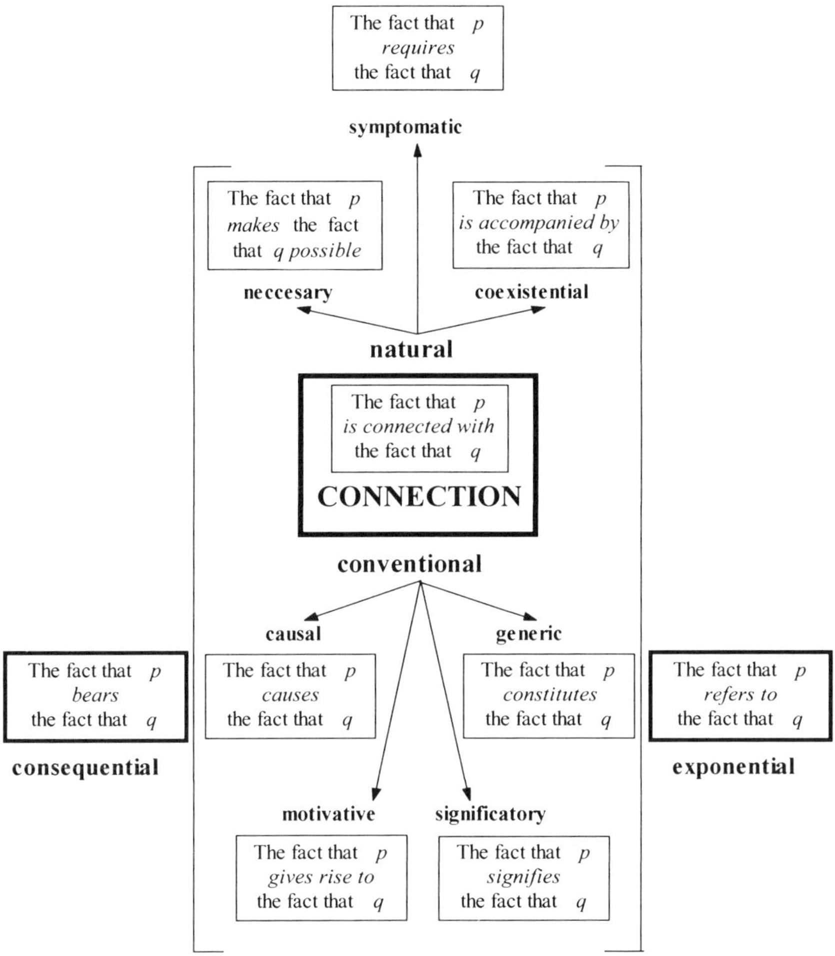

Fig. 4. Types of connections presupposed by conditionals

9.8

As can be seen, in contrast with Ajdukiewicz, in taking the holding of a certain type of connection to be a component of the objective (referential) content that differentiates various conditional-types from one another (i.e. taking it to be a component of the set of states of affairs stated by such a sentence), I do not have to regard conditionals of the type "If the moon is made of green cheese, I shall die on a day the date of which is an even number", since these states of affairs are not connected in any of the above-mentioned ways.

Stanosz, a proponent of Ajdukiewicz's approach, writes what follows about the view according to which the reason for denying that a conditional of the form "If *p* then *q*" is true may be a lack of connection between the fact that *p* and the fact that *q*:

> In order for [...] this view to deserve being called a solution to the problem of th truth-conditions for conditionals, the concept of a content connection between the antecedent and the consequent needs clarifying. [...] Yet all hitherto proposed definitions of such a concept are clearly inadequate in that they correspond at most only to certain particular senses of the conjunction *if ... then* [Stanosz 1985, p. 75].

I cannot agree that a disjunctive definition of a protocolar kind would be inadequate, quite simply because the *definiendum* is itself polysemic. Neither do I agree that, in the case of conditionals like the following:

If Jan obtains a passport, he will leave the country,

"the concept of a content connection between the constituents of acceptable conditionals defies any attempt at explanation" [Stanosz 1985, p. 75]. It is true that here there is no causal connection but *there is* a connection, which I have called a "motivational connection" above. In contrast to Stanosz, I hold that the consequent of the above conditional can be inferred from its antecedent, namely as its enthymematic consequent on the basis of, e.g., "a generally accepted truth": "If someone does not obtain a passport, s/he will not (legally) leave the country, even if s/he wanted to. Jan wants to leave the country, therefore ...".

It is another matter entirely that *this inferrability* is not stated by the relevant conditionals. So those (for example [Grodziński 1969, p. 64] and [Wolter and Lipczyńska 1973, p. 98]), who hold that a sentence of the form "If *p* then *q*" states that *sentence* "*p*" implies *sentence* "*q*" are wrong. Of course, *if* a sentence of the form "If *p* then *q*" is true — and thus *states*, amongst other things, that *q* (i.e. the state of affairs indicated by "*q*") is connected with the fact that *p* (i.e. the state of affairs indicated by "*p*") — *then*, obviously, the antecedent implies the consequent.

N.B. A proof that a conditional like "If he comes, I shall give him a piece of my mind" presupposes something more than material implication is given by the fact that contraposition would not be acceptable in the case of a natural language conditional of this type. The sentence "If I don't give him a piece of my mind, he won't come" has a completely different referential (objective) content from that of its contrapositive, for it presupposes that giving somebody a piece of my mind leads to his not coming.

9.9

(14) Material implication "$p \rightarrow q$" presupposes, amongst other things, that it is not the case that p unless it is the case that q.

(15) Every conditional of the form "If p then q" implies the material implication "$p \rightarrow q$".[2]

Like Stanosz, I believe that it is a significant virtue of a theory that it allows the application of logic (in this case, propositional logic) to the analysis of natural language. In any case, like Pelc, I am already satisfied if logical theses can be used as auxiliary negative criteria in discovering the truth. And I hope that the solution to the problem of conditionals outlined above may be seen as following this path.

(16) A conditional of the form "Supposing that p, q" expresses, amongst other things:
(a) ignorance (on the part of the speaker) as to whether p;
(b) ignorance (on the part of the speaker) as to whether q;

---

[2] Perhaps W.v.O. Quine also had this in mind when he wrote:

> Usage conforms to the third line of the table (*VF*) and usage lapses as soon as a case is precisely located elsewhere in the table [Quine 1940: 23]. Thus it is that the indicative conditional of ordinary language has given way in the logically regimented language of science to the material conditional, which while serving still the scientific purposes of the old, does not share the defectiveness of the old with regard to truth-values. The material conditional formed from any two statements has a definite truth-value; discovery of the falsity of the antecedent of a material conditional disposes of the questions of the truth-value not by dismissing it, but by delivering the answer "true" [Quine 1953b, p. 165].

One cannot, however, exclude the possibility that he had in mind a greater independence between conditionals and material implication, since he holds that in the case of the other natural languages conditionals also are *idle or senseless* [Quine 1940, p. 23].

(c) readiness (on the part of the speaker) to infer the sentence "*q*" from the sentence "*p*".

N.B. One should bear in mind that ignorance as to whether *p* is also expressed by the question "Is it the case that *p*?".

9.10

I am aware of the fact that the solution presented above is not complete. In particular, it presupposes the solution of certain basic philosophical problems, namely:

(a) a clarification of the concept of "rationality" (cf. thesis (1));

(b) an indication of the criteria of identity and distinctness of sentences (cf. thesis (4) and thesis (5));

(c) an analysis of the concept of "content" as the set of states of affairs referred to by a given sentence (cf. thesis (10));

(d) a description of the criteria for each particular type of connection presupposed by conditionals (cf. thesis (13)).

Some of these problems have been dealt with in the works of Polish logicians (cf. [Jadacki 1980]), other have been touched upon in some of my own articles [Jadacki 1981; 1982], but the last word on the matter has undoubtedly yet to be written.

10. ON NONSENSE

10.1

Thomas G. Winner began his paper delivered during the Semiotic Symposium at Radziejowice (near Warsaw) with the following words:

> Prague, in the inter-war years, was a focal point of the European avant-garde [...] [Winner 1984, p. 457].

One of tricks serving the then avant-garde to liberate art from the constraints of "the objective world" was to use various types of nonsense in their pieces. At the same time, the opposite tendency appeared in the philosophy and science: to get rid of everything that could be suspected of being nonsense.

It is worth noting that appealing to the «sense of nonsense» is something peculiar to the authors trying to convince us of the reductionistic programme (e.g. reists very often talk about nonsense). The word "nonsense" or its synonyms are used at least 14 times in George Berkeley's *Treatise*!

I think all this is sufficient to justify an attempt to explicate the notion of "nonsense".

10.2

When confronted with a sentence *S* a given user *U* can take one of two assertive attitudes: s/he can either accept or reject *S*. Let us assume that the following equivalence holds:

(1) *U* rejects *S* *iff* *U* accepts the negation of *S*.

Thus, from now on, we can speak about acceptance instead of assertive attitudes.

Acceptance is subordinated to two duties. On the one hand:

(2) If *U* accepts *S* then *S* should be JUSTIFIED.

Let us notice that the following dependence holds:

(3) If *S* is justified then *S* is JUSTIFIABLE.

If so, then:

(4) If *U* accepts *S* then *S* should be JUSTIFIABLE.

On the other hand:

(5) If *U* accepts *S* then *U* should comprehend *S*.

The duty expressed by formula (5) can be equivalently expressed as:

(6) If *U* accepts *S* then *S* should be comprehensible to *U*.

Formula (6) entails:

(7) If *U* accepts *S* then *S* should be comprehensible.

The following dependence holds:

(8) If *U* is comprehensible then *S* is meaningful.

If so, then:

(9) If *U* accepts *S* then *S* should be meaningful.

This dependence was already formulated by Berkeley. He writes: "In order to avoid absurdity and scepticism, I should [...] be obliged to give up my opinion" [Berkeley 1714, p. 172]. He links also *expressis verbis* incomprehensibility with nonsensicality, writing that "to assert that which is inconceivable, is to talk nonsense" [Berkeley 1714, p. 215].

Sentences that do not fulfil the conditions given in formulae (2), (5), (7) and (9), i.e., groundless, unjustifiable, incomprehensible, and meaningless sentences, constitute various kinds of nonsense.

I would like to note, by the way, that, in Poland in the inter-war period, a very interesting discussion concerning comprehensibility took place. It was opened by Karol Irzykowski, and continued i.a. by Jerzy Hulewicz [1924], Józef Ujejski [1924], and Stanisław Ignacy Witkiewicz [1927]. For instance, Hulewicz noticed, referring to Cyprian Kamil

Norwid, that "something is comprehensible when it should be (not necessarily is) comprehende by someone".

We are inclined to say that a person who accepts nonsensical sentences, i.e. sentences that are groundless, unjustifiable, incomprehensible, and meaningless, behaves IRRATIONALLY.

In this way, the problems of nonsense and (ir)rationality interlock.

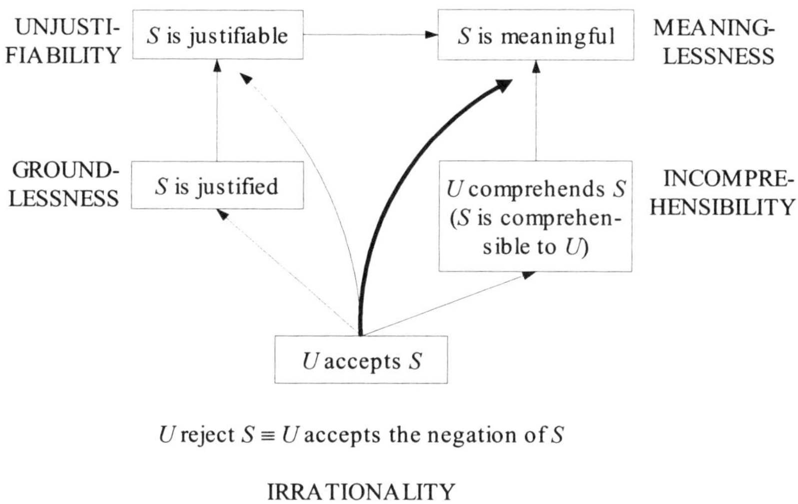

Fig. 5. Nonsense and irrationality

10.3

These two paths of dependencies — one leading from acceptance to justifiability, the other leading from acceptance to meaningfulness — meet:

(10) If *S* is justifiable then *S* is meaningful.

In this way, meaningfulness of *S* appears to be a necessary as well as a «key» condition of the acceptance of *S*: it is a «key» condition because both paths lead to it: *via* justifiability and *via* comprehensibility.

It is disputable whether the justifiability of *S* is equally crucial or not: whether or not converse dependence to the described in formula (10) is in force here:

(11) If *S* is meaningful then *S* is justifiable.

In other words, it is disputable whether:

(12) *S* is meaningful *iff S* is justifiable.

As is well known, neopositivists answered this question affirmatively, imposing here an additional condition:

(13) *S* is justifiable *iff S* is EMPIRICALLY verifiable.

By this means, their decision adopted the following form:

(14) *S* is meaningful *iff S* is EMPIRICALLY verifiable.

This decision was not free from difficulties, which became the object of penetrating analyses done by the strongest minds during several decennaries. The question of whether or not meaningful expressions can stand in logical relations with nonsense, and if so, what are the consequences of this fact, appeared to be a matter of special complexity. One of the proposed solutions was to consider nonsense as the third logical value, beside truth and falsehood. It made an affirmative answer to the above question possible, but it did so at the price of modifying the sense of the functors of the classical propositional calculus.

A very clear report of this discussion is given by Charles Frederick Presley [1961].

10.4

When analysing the usage of the word "nonsense" one notices easily that it is a kind of epithet: that nonsense is something improper. In the light of the dependencies mentioned above, it can be explained by indicating that acceptance of a nonsensical sentence is improper because of the fact that it violates a certain duty binding on us, if our behaviour is to be rational. If the *U*'s behaviour is to be rational then we expect that S accepted by *U* is not nonsense.

Tadeusz Tomaszewski was one of the first persons who noticed that the necessary condition of considering something as a nonsense are certain expectations. As a matter of fact, his thesis is much more radical:

> The assessment of the nonsensicality of a text depends on reader's expectations. The text itself has no objective features that qualifes it as absolutely nonsensical [Tomaszewski 1936, p. 38].

(I do not agree with the opinion expressed in the second sentence of this quotation.)

The explication of "nonsense" should start with the remark that this epithet is used with reference not only to sentences — and even not only to expressions — but also to extralinguistic objects.

It is true with reference to every piece of nonsense that:

(15) If *x* is nonsense then (a) *x* lacks something or (b) something in *x* is replaced by something «improper».

Let us agree to use the word "asense" in the case of nonsenses that fulfil only condition (a), and the word "countersense" in the case of nonsenses that fulfil only condition (b).

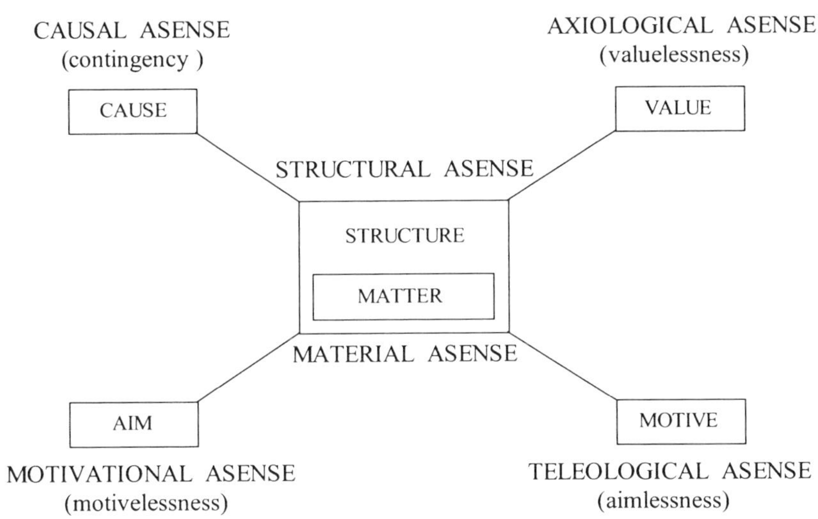

Fig. 6. Types of asenses

The «things» that are lacking (because of the lack of which a given object is an asense) are either parts of this object or inner bonds between its parts, its cause or motive, its value or aim. Thus, asense can be alternately material, structural, causal, motivational, axiological and teleological.

Countersenses have their main source either in exchanging a certain part of a given object for a part of a different («improper») kind, or in exchanging a determined bond between parts of a given object for a different (also «improper») kind. These are alternately material and structural countersenses.

In all these cases, we call something "nonsense" if we expect that this "something" should have some property:

(a) in the case of material nonsense, we expect that a given object should have determined parts, while either one of them is *de facto* missing (asense) or it is changed for the «improper» (countersense);

(b) in the case of structural nonsense, we expect that parts of a given object are bound in a determined way but we either find no bond between them (asense) or the existing bond is the «improper» part (countersense);

(c) in the case of causal asense, we expect that a given object, in particular, a certain event, has a determined cause but it turns out that this event is «accidental»;

(d) in the case of motivational asense, we expect that a given object, in particular, a certain action, has a determined motive but it is, as a matter of fact, motiveless;

(e) in the case of axiological asense, we expect that a given object has determined value, whereas it is valueless; Carl Gustav Jung narrows the notion of "nonsense" in this way by saying that "*das geistige Pendel schwingt zwischen Sinn und Unsinn, und nich zwischen richtig und unrichtig*" [Jung 1961; p. 158];

(f) in the case of teleological asense, we expect that a given object, in particular, a certain action, has a determined aim but it is in fact aimless.

It seems that there is also a kind of asense, which can be called "probabilistic asense", and which is similar to causal and motivational asenses. We can speak about probabilistic asense if we expect that a given object (in particular, an event) occurs with a certain probability but our expectation turns out to be incorrect.

10.5

All the types of nonsense mentioned above can be also encountered in the domain of language. It concerns, in particular, material and structural nonsenses, which, in the case of expressions, we call "lexical" and "syntactical" nonsenses respectively.

The examples of lexical nonsense are *Słopiewnie* by Julian Tuwim. Here you have the beginning of the poem "Słowisień":

> W białodrzewiu jaśnie dźni słoneczno,
> Młodzie złoci białopałem żyśnie,
> Drzewia pełni pszczelą i pasieczną,
> A przez liście kraśnie pęk słowiśnie [Tuwim 1923, p. 144].

Roman Ingarden was the philosopher who called our attention to *Słopiewnie* and the so-called "mirohłady" in this context [Ingarden 1934, p. 4].

In the case of syntactical nonsense, one can, following Tadeusz Kotarbiński, speak either about nonsense *tout court* or syntactical inconsistency or syntactic disorder [Kotarbiński 1929, p. 20]. The fact that our expectations are not fulfilled in this respect is determined by the syntactic rules of a given language. Thus Witold Marciszewski writes:

> A system of expressions *U* is nonsense (with respect to a language *J*), when it is a chain of expressions built in violation of the syntactic rules of *J* [Marciszewski 1970d, p. 190].

One can try to explicate lexical nonsense of compound expressions by identifying nonsensical members of these expressions (possibly also functors) with variables of a proper category.

Let us consider the Polish expression:

(i) Młodzie złoci białopalem,

which is lexical nonsense with respect to its first and third member. Let us agree that this expression has the following semantic and syntactic structure (written in Kazimierz Ajdukiewicz's notation):

(ii) $n\ s/nn\ n$;

(iii) (1,1), (1,0), (1,2).

If so, then (i) is equivalent to the function:

(iv) x złoci y.

If we give (i) the form of the expression:

(v) Młodzie zboci białopalem,

we shall receive the function:

(vi) *Rxy*,

that is (v) will be equivalent to the fact that a certain object stands in a certain relation to a certain object.

10.6

I have already mentioned that nonsense may have its source in the lack of justification; one can call it "argumentational nonsense". In addition to argumentational, lexical, and syntactical nonsenses we have *sui generis* linguistic nonsenses, associated with the special functions fulfilled by linguistic expressions.

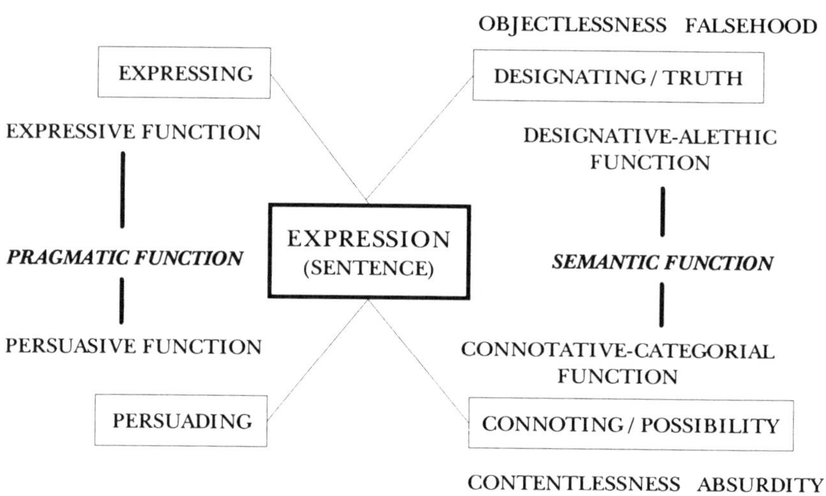

Fig. 7. Semiotic asenses

I mean the following functions:

(a) connotative function: expressions can CONNOTE certain properties;

(b) designative function: expressions can DESIGNATE certain objects;

(c) expressive function: expressions can EXPRESS certain experiences of the sender (her/his beliefs in particular);

(d) persuasive function: expressions can PERSUADE the receiver to a certain behaviour (a linguistic behaviour in particular).

In the case of sentential expressions, the connotative (a) and the designative (b) function are replaced by new functions that might be called "categorial" and "alethic". A given sentence satisfies the categorial function if it describes a POSSIBLE state of affairs. A sentence satisfies the alethic tunction if it is TRUE, i.e. if it describes an actual state of affairs.

Thus, *sui generis* linguistic asenses consist in the fact that at least one of the following functions is not satisfied:

(a) connotative-categorial — when a given expression is «contentless», or ABSURD in particular, i.e. a sentence "*p*", for which the formula "It is impossible that *p*" is true;

(b) designative-alethic — when a given expression is «objectless» (empty), or FALSE in particular, i.e. a sentence "*p*" for which the formula "It is not the case that *p*" is true;

(c) expressive;

(d) persuasive.

Now we shall notice that claiming that "*Tautologie und Kontradiktion sind aber unsinnig; sie gehören zum Symbolismus*" [Wittgenstein 1922, p. 98], Ludwig Wittgenstein means lexical-syntactical nonsense by "*unsinnig*". Claiming that "*Tautologie und Kontradiktion sind sinloss*", because "*ich weiß z.B. nichts über das Wetter, wenn ich weiß daß es regnet oder nicht regnet*" [Wittgenstein 1922, p. 98], he understands rather connotative-categorial nonsense by "*sinnloss*" (see below). The same is true when he claims that ""*Sokrates ist identisch" heißt darum nicht, weil es keine Eigenschaft gibt, die "identisch" heißt*" [Wittgenstein 1922, pp. 126—128].

Some people (e.g. Annette C. Baier [1967, p. 520]) do not want to treat falsehood (not even the so-called evident falsehood said «in the face of» a respective fact) as nonsense. The fact that in the case of some expressions (e.g. "Virtue is green") we hesitate whether it is falsehood or (categorial) nonsense suggests surely that such a position is not correct.

As a starting point for explicating the notion of "categorial meaningfulness", we can take the following formula:

(16) If the expression "*Pa*" is categorially meaningful then for a certain *Q* ⟨it is not the case that for every *x* (*Qx*) and {for every *x* [(if *Px* then *Qx*) and *Qa*]}⟩.

It seems that this interpretation renders the idea of presupposition as a sentence (or belief) of such a kind that its truth is necessary to the meaningfulness of a given expression. The expression "Virtue is green" is a nonsense because the presupposition "Virtue has a colour" is a falsehood. It seems generally that the following equivalence is true:

(17) α is a presupposition of a sentence *A iff* if *A* claims that *p*, and α claims that *q*, then if *q*, then it is possible that *p*.

One can say that a given expression is COMPLETE NONSENSE when it fulfils no function from among the ones mentioned above. The «material» of such complete nonsenses would determine, whether they are still expressions or some extra-linguistic objects: a kind of visual art or a musical piece. On such an approach:

(18) *E* is an expression of a language *L iff E* is a chain of sounds (or letters) belonging to the articulatory base (or the alphabet) of *L*.

If we expect that a certain chain of sounds or letters is an expression in this sense but id does not *de facto* fulfil the condition indicated by formula (16), then this chain is an ARTICULATIVE-CALLIGRAPHIC NONSENSE. It would be GIBBERISH *sensu stricto* (cf. [Jadacki 1985, pp. 251ff]).

10.7

The connotative-categorial nonsenses can be of various types, depending on which kind of impossibility is their source. We should take into consideration here:

(a) theoretical impossibility, i.e. logical and ontic (physical) impossibility;

(b) practical impossibility.

N.B. Following Kotarbiński, it has been generally agreed to use the term "absurd" only as signifying logical absurd, i.e. either a sentence that is internally contradictory [Kotarbiński 1929, p. 57], which can have either of two forms:

(a) $K\, p\, N\, p$,

or of the form:

(b) $(Ex)\, K\, (Px)\, N\, (Px)$,

or a sentence implying sentences of the form (a) or (b). One can put "*N* It is possible that", of course, before (a) as well as before (b).

The expressions coming into being, when — as Thomas Hobbes writes:

> I ascribe to the giving of names of *bodies*, to *accidents*; or of *accidents*, to *bodies*. As they do, that say, *Faith is infused*, or *inspired*; when nothing can be *powered*, or *breathed* into any thing, but body [Hobbes 1651, p. 35],

belong to the kind of physical nonsense.

The examples of this kind of nonsense can be found, i.a. in Stéphane Mallarmé's poems. Here you have the beginning of his poem "Cantique de saint Jean":

> Le solei que sa haite
> Surnaturelle exaite
> Ausitôt redescend
> Incadnescent ... [Mallarmé 1887: 495].

Now we add that:

(19) *S* is theoretically impossible *iff S* is inconsistent (contradictory including) with laws of a certain (logical or physical) theory;

(20) *S* is practically impossible *iff S* describes a state of affairs that cannot be reached by available means.

It is evident at once that because new (sometimes competing) logical as well as physical theories appear and because technical «abilities» change, the pool of nonsense also changes. Berkeley notes in this context that "there was a time when the antipodes and motion of the earth were looked upon as monstrous absurdities even by men of learning" [Berkeley 1714, p. 64].

In a logical system that does not accept the principle of non-contradiction, neither the negation of this principle nor its consequences are, of course, absurd.

The instability of limits of nonsense is also evident in the domain of natural language, the theoretical background of which consists in common sense. This is just one of the reasons for the fact that, as Leon

Chwistek writes, "in every-day language the transition between sense and nonsense is not sharp" [Chwistek 1935, p. 37]. The other reasons include the fact that neither lexical base nor syntactic rules of natural language are determined precisely.

10.8

The fact that we recognise a given expression, which does not fulfil a certain function as nonsense of a respective type, depends on our expectations with regard to this expression.

It is worth noting that in the case when our expectations are not fulfilled, recognising them as nonsense is not the only rational strategy.

In such a situation, we can also:

(a) try to reconstruct the missing element in LITERAL NONSENSE (and to find the missing or deformed function) with the aid of suitable INTERPRETATIVE operations;

(b) decide that in a given case what appears to be an expression is no expression at all (or that it dose not fulfil a certain function), and so that the expectation of a (determined) linguistic function is pointless.

If we choose the first strategy then in the case of expression (i) "Młodzie złoci białopalem" the weakest interpretation of the proper function (iv) would consist in binding its variables with the particular[1] quantifier:

(vii) (*Ex*) (*Ey*) (*x* złoci *y*).

In the case of the expression (vi) "Młodzie zboci białopalem" such an interpretation would be:

(ix) (*Ex*) (*Ey*) (*ER*) (*Rxy*).

Ajdukiewicz chooses the second strategy when he says that the chain of letters "Every let us after is a man" is an expression but not a sentence [Ajdukiewicz 1934d: 97].

---

[1] According to the main representatives of the Lvov-Warsaw School, existential quantifiers have nothing to do with the problem of ontological commitment of a given theory; this fact was stressed by using the term "particular quanitifier" or simply "particualisator".

Both these strategies were mentioned by Tomaszewski [1936, pp. 20ff].

The type of our expectations is determined by many things. In the case of texts, but also in the case of paintings, the decisive factor is often either the title or the subtitle of a given text or a given painting. The text described by its author as a fable or a phantastic novel is read by us with quite a different attitude from the attitude we have when reading philosophical essays or scientific dissertations.

10.9

It seems that neither in (analytical) philosophy nor in science, do we expect that expressions used fulfil the function of expressing or persuading; thus the problem of expressive or persuasive nonsense does not arise there.

For that reason, perhaps, Hobbes was right in considering the creation of nonsense as a privilege "to which no living creature is subject, but man only". But he certainly was not right in adding that "of men those are of all most subject to it, that professe philosophy" [Hobbes 1651, p. 35].

One of the greatest «containers» of nonsense is art (*cf.* surrealism) and literature (*cf.* pure nonsense and grotesque). The world presented in Marie Germinova's *A sleeping woman* (London) is material asense, in Henri Matisse's *Inanimate nature with magnolia* (Musée National de l'Art Moderne, Paris) — structural asense, in René Magritte's *The human law* (Tokyo) — structural countersense. Władysław Strzemiński's *Composition* (1947—1949; Muzeum Sztuki in Łódź) could be an example of lexical asense.

Leopold Staff wrote very rightly about Konstanty Ildefons Gałczyński's poetry:

> Pokazałeś w wesołej herezji
> Przez swe fraszki fiołkowe i gęsie,
> Ile jest nonsensu w poezji
> I ile poezji w nonsensie [Staff 1958].

A good example is here Ladislav Fuks' *Myši Natalie Mooshabrové* [Natalia Mooshaber's Mice], analysed in detail by Winner in his paper cited above. As he wrote in his conclusion:

> In Fuks' world view [...] a conventional plot is only weakly suggested and serves to emphasise the absurd, nonrational world in which time moves forward and backward, individuals merge into each other and into the outside world; the living and the dead merge, the dead interact with the living [Winner 1984, p. 476].

Nonsense in philosophy, and in science, can take one of four forms: argumentational nonsense, lexical-syntactical nonsense, absurdity and falsehood.

Escaping the nonsense of the analysed kinds is what I mean by "responsibility for the word".

PART III. UNDERSTANDING AND SILENCE

11. ON MISUNDERSTANDINGS ABOUT UNDERSTANDING

11.1

There are two misunderstanding about the relation of understanding. The first misunderstanding is that the necessary condition of defining (and characterising) *understanding* is to indicate the field (i.e. the domain and the counterdomain) of the relation of understanding. The second misunderstanding is that, among others, unobservable objects (for example, intentional objects) belong to the counterdomain of this relation.

It is also a fact that the domain of the relation of understanding contains people (N.B. certainly not all people), animals, and perhaps computers, whereas the counterdomain (or one of the counterdomains at least) is created by observable objects.

11.2

As to the semantic category of "understanding", the misunderstanding consist in the view that the word "understand" depending on a given context has one of the following categories: *s/nn*, *s/ns*, or *s/nnn*. As a matter of fact, these categories represent of three different functors:

(i) *s/nn* — *x* understands *y*,

(ii) *s/ns* — *x* understands that *p*,

(iii) *s/nnn* — *x* understands *z* by *y*.

Each of these functors refers to a certain two- or three-place relation. It is a complete list, apart from the metaphorical contexts (for example, of the kind of Martin Heidegger's declarations that man ... is understanding).

11.3

"Understanding" is a double polysem. Firstly, the polysemy peculiar to every mental predicate characterises this word. "Understanding" as a mental predicate can refer either to a certain disposition (*habitual* understanding), to a certain act (*actual* understanding), or to a certain chain of acts (*processual* understanding). Secondly, "understanding" is also polysemic in its own unique way. One can call it "four *i*-s polysemy". For to understand is either to feel (*intuitive* understanding), to be conscious (*identificatory* understanding), to justify (*indulgentive* understanding), or to signify (*inscriptive* understanding).

It is a misunderstanding to take, that the habitual, actual, and processual understanding, on the one hand, and the intuitive, identificatory, indulgentive and inscriptive understanding, on the other hand, are members of two logical classifications of the denotation of "understanding". These are various MEANINGS of "understanding" but not various subsets of one set of understandings.

Let us add that intuitive understanding (i.e. feeling), which include empathy into others' feelings as well as insight into the essences of things, is a kind of cognition. Being conscious of something can be described as identificatory understanding only if the object we are conscious of is either somehow hidden from our sight or at least hard to realise on a certain account. Indulgentive understanding — motivative or motivative-behavioural one — is of various degrees: from approving and recognising as right (motives or behaviours), throughout consenting and permitting, until non-damnation and excusing. Finally, inscriptive understanding is a three-place relation and can be a connotative or referentive understanding depending on what is signified by something: the *connotatum* or the «referent».

11.4

The term "identificatory understanding" appears in various elliptic contexts. The phrase "*x* understands *y*" can mean that:

(a) *x* is conscious of the nature of *y* (*cardinal* understanding);

(b) *x* is conscious of the structure of *y* (*constitutional* understanding) ;

(c) *x* is conscious of the horizon of *y* (*contextual* understanding);

(d) *x* is conscious of the reason of *y* (*causal* understanding: *initial* one — of causes or motives, or *effectual* one — of effects or aims);
(e) *x* is conscious of the type of *y* (*categorial* understanding);
(f) *x* is conscious of the sense of *y* (*conceptual* understanding).

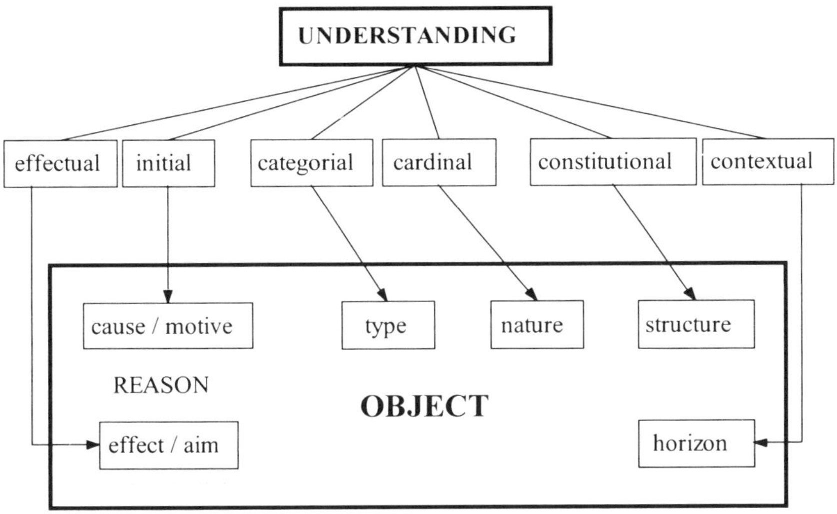

Fig. 8. Ellipticity of "understanding" in the identificatory sense

This ellipticity of six *c*-s is probably derivative of the polysemy of "sense" *sensu largo* ("nature", "structure", "horizon", "reason", and "sense" *sensu stricto*).

However, it would be a misunderstanding to think that cardinal, constitutional (syntactical), contextual (psycho-contextual and psycho-contextual-behavioural), causal (psycho-causal and psycho-causal-behavioural), categorial and conceptual (verbal) understandings are all types of understanding. For that reason, it is also a misunderstanding to speak (literally) about WAYS or METHODS of understanding.

11.5

Each of the current uses of "understanding" (and the derivates of this word) can be paraphrased with the aid of the specialised terms introduced here. However, it would be a misunderstanding to require the paraphrase of every current use of "understanding" by means of EXACTLY ONE of these specialised terms.

Let us consider the following context:

> We are not downhearted. The only trouble is, we cannot understand what is happening to our neighbours [Chamberlain 1906].

The sense of "understanding" oscillates here between the identificative and indulgentive ones. Similarly, it would be pointless to engage in the controversy surrounding the question whether to understand music is to be conscious of the (formal) structure of a given work, or of the mental acts of the composer that accompany the creation of this work, or of the motives of composing, or perhaps something altogether different (for example, to react properly to this work in the emotional sphere).

11.6

The diversity of opinions on the matter of understanding either has its source in the definition of meaning (i.e. some of these opinions are misunderstandings) or these opinions are empirically testable (i.e. some of them are simply false).

A good example of the first situation is the SOPHISM OF THE HERMENEUTIC CIRCLE. In order to understand, one already has to understand. In particular, to know *why* somebody does something, one should know *that* somebody does this thing — and these are to be facts, which cannot be identified separately. The press of the button by a certain person is to be (causally) understood only by the man, who knows that this person wanted to press the button. In fact, it is understood by the man, who knows that this person wanted (for instance) to ring. In general, it is not the case that:

If we understand a given activity of a certain person, we know that this person wanted to carry out just *this* activity.

A good example of the second situation is admitting (ungrounded) to the QUESTION OF THE CONGENIALITY OF DISPOSITIONS TO UNDERSTAND the status of the philosophical (and not empirical) problem.

11.7

There are four main kinds of misunderstanding concerning the relations between the various notions of "understanding".

The first consists in the view that certain objects can be understood only by means of an *intuitive* understanding. In fact, however, we do not know any results of this intuitive understanding that could not be reached with the aid of «normal» methods (e.g. reasoning). Of course, this is not a sufficient reason for rejecting the existence of such understanding. But IS it arguable at all that something does not exist?

Second, tt is a misunderstanding to consider the notion of "*identificative* understanding" or, more precisely, of "conceptual understanding" as the generically primitive notion.

The third misunderstanding concerns the view that a necessary condition of the indentificative understanding of a certain object is the *indulgentive* understanding of this object. Let us compare the following declaration:

> All, everything that I understand, I understand only because I love [Tolstoy 1865—1869].

An analogically baseless supposition is the view that a necessary condition of the psycho-contextual understanding is to *feel* (or to have the ability to feel) the proper («understood») experiences. It is even more baseless to believe that one can (conceptually) understand the utterance of a certain person *better* than this person her/himself. (It was said about Tadeusz Kotarbiński that his summaries of others' papers showed that he *understood* the utterances of the authors of summarised papers *better* that the authors themselves.)

In the extreme, the idea would be that it is possible to understand (conceptually) an utterance that is incomprehensible to the sender. It should be added, of course, that I am concerned here with the *proper* conceptual understanding (i.e. the reconstruction of the sense meant by the sender) and not with just *any* understanding (i.e. ascribing any sense).

Finally, the fourth misunderstanding concerns the so-called THE PRINCIPLE OF HERMENEUTICS, according to which the necessary condition of *inscriptive* and conceptual understanding is to reach the full contextual understanding (of the interpreted utterance). It is supposed, moreover, that it is just impossible to attain such a full contextual understanding.

11.8

We have the following dependencies between understanding and other mental functions:

(1) If somebody understands something identifically (in the actual sense), he THINKS.

(2) If somebody understands the utterance of a certain person conceptually (in the actual sense), he PERCEIVES this utterance.

Of course, this has to do with the «understanding» perception, and not with the empty «gaping».

(3) If somebody has the ability to SPEAK, he has also the ability to understand conceptually.

We are talking here about speaking, not about «babbling».

(4) If somebody INFORMS a certain person of something by means of a certain utterance, both of them understand this utterance conceptually.

(5) If somebody understands the utterance of a certain person conceptually, this person EXPRESSES something by means of this utterance.

N.B. It may happen that the thing expressed differs principally from the thing understood. Here we have the source of the remark:

> Speech was given to man to disguise his thought [Talleyrand ca. 1800].

(6) If somebody REACTS to a certain work of art EMOTIONALLY in an appropriate way, he understands it conceptually.

This dependence should be certainly limited to the domain of the works of literary art.

(7) If somebody understands inscriptively by a certain expression something determinate, he CONCEDES the statement that this determinate thing is the sense of this expression.

(8) If a certain person understands inscriptively (in the habitual sense) by a certain sign a certain object, the presentation of this object for this person is ASSOCIATED with the presentation of this sign.

On the other hand, it is a misunderstanding to consider understanding as a kind of REASONING. The source of this misunderstanding is probably the confusion of understanding with the INTERPRETATION, i.e. coming to understanding. There are many techniques of achieving this state: expressive, aesthetic, symbolic, hermeneutic, psychological, humanistic, imaginative, teleological, psychoanalytical, and — last but not least — logical interpretation. In contrast to interpretation, understanding (except for intuitive understanding) is either only a certain (potential or actual) state or a chain of cognitive states.

It is also a misunderstanding to think that we can consider something (i.e. a certain sentence) as true and not understand it at the same time. Put in a nutshell:

> The ignorant man always adores what he cannot understand [Lombroso 1864].

Let us add that it is hard to imagine a more essential misunderstanding than questioning the dependence:

(9) If somebody understands something conceptually, the thing understood is a SENSEFUL expression.

On the other hand, it is an open question whether one can say internally contradictory utterances about conceptual understanding.

11.9

Many facts speak for the view that identificative understanding (or being conscious of something) admits of degrees. I think the following dependencies obtain:

(1) The more detailed the structure, the wider the horizon, and the more distant the reasons of an understood object a given person is conscious of, the *better* this person *understands* indetificatively this object.

(2) The more elements of compound expression understood conceptually, the *better* this person *understands* this expression.

(3) Let us suppose that:

(a) somebody understands a certain expression *selectively*, if s/he is aware only of a part of the denotation of this expression;

(b) somebody understands a certain expression *clearly*, if s/he is aware of the whole denotation of this expression;

(c) somebody understands a certain expression *distinctly*, if s/he is aware of any connotation of this expression;

(d) somebody understands a certain expression *accurately*, if s/he is aware of the lexical connotation of this expression;

(e) somebody understands a certain expression *exactly*, if s/he is aware of the essential connotation of this expression.

Now, the closer somebody's conceptual understanding of a given expression on the scale from (a) to (e) is to the bottom (i.e. to exact understanding), the *better* he understands this expression.

(4) The more detailed presentation is being conscious of the sense of a given expression, the *better* the understanding person *understands* (conceptually) this expression.

On the other hand, it is a misunderstanding to claim that the greater the distance between the literary sense of a given expression and the sense of which a certain person is aware, the *better* this person understands this expression (conceptually).

11.10

It would be a misunderstanding to require the reconstruction of the normal, equivalent definition of "proper conceptual understanding" (or "proper comprehending"). We have only:

> If the object *z* is understood inscriptively by the sign *y*, then the person *x understands y properly iff x* understands inscriptively by *y* (the object) *z*.

It is another matter that the initial condition of this definition is not precise. Does "It is understood" mean: "Everybody understands",

"Most people understand", "Somebody understands", or "(Each of, most of, or some of) proficient users understand"?

The normal definition can be given only for "some understanding":

> The person *x* has *some understanding* of the sign *y iff x* understands (inscriptively) something by *y*.

But it would be a misunderstanding to identify proper comprehending with any comprehending (i.e. with *instrumental* comprehending). And the view that instrumental comprehending of a certain object is identical with being aware of the fact that this object is a sign — is an even greater misunderstanding.

11.11

The basic misunderstanding is the belief in the existence of universal criterion of comprehending and understanding in general — a criterion that could be a necessary as well as a sufficient condition. It is equally a misunderstanding to look for any criterion *sensu stricto*, i.e. an absolute criterion. In fact, all we have at our disposal are criteria *sensu largo*, i.e. partial criteria:

(a) the *feeling* of a correct understanding of a given utterance;
(b) correct *reproducing* (i.e. quoting) of a given utterance;
(c) correct *explaining* (the sense of) of a given expression;
(d) correct *summarising* of a given utterance.

The above criteria are merely partial criteria because of the following facts.

The source of a feeling of misunderstanding at a given utterance can be found in disturbances of the attention of the complexity of the object described in this utterance. The source of *mistakes* and difficulties in reproducing, explaining, and summarising may lie in defects of intellectual dispositions other than dispositions to understand (for example, problems with memory, mistakes in reasoning), or even purely articulatory defects. *Improper* reactions to a given utterance can arise from gaps in extra-linguistic knowledge or disfavour to take a hoped-for stand (for example, in the case of a lying answer).

On the other hand, the correctness of the respective behaviours can be a result of incidental coexistence. One can find the proper *definiens* without understanding it. It is very hard to make sure whether a given utterance *is* the answer to the question asked, and whether this utterance bears witness to understanding this — and not another — element of the formulated question.

11.12

What does possessing linguistic competence mean?

It consists of two abilities: the ability of conceptual *understanding* utterances already heard, and the ability of *producing* (or complementing) utterance — among others, (composed) utterances never heard before.

> What we genuinely understand, we can do [Williams 1958].

The main misunderstanding here is the view that the necessary and the sufficient condition of such an ability is the understanding of the structure and the sense of the elements of such utterances. The existence of idioms (understood *en bloc*) and of the ability to create complex expression by means of syncategoremata (very often not understood in isolation) speaks for the faultiness of the view mentioned above.

12. ON DEFINITION, EXPLICATION, AND PARAPHRASE

12.1

To begin with, let me make six assumptions, which, I hope, are not false, about each of my Readers.

First, I assume that my Reader knows all Kazimierz Ajdukiewicz's works, or at least those papers that concern *definitions* and where his *method of paraphrase* is either described or applied [Ajdukiewicz 1965b; 1977].

Second, I assume is that my Reader knows Kazimierz Twardowski's distinction between *operations*, *results* and *aims* of operations [Twardowski 1894].

Third, I assume that my Reader knows Jan Łukasiewicz's distinction between what is *given* and what is *asserted* which underlies his classification of reasoning: [Łukasiewicz 1970].

Fourth, I assume that my Reader accepts the *semiotic triad*, i.e. the distinction between three aspects of a symbol:

— its *form*,

— its *connotation* (i.e. a sequence of characteristic properties),

and

— its *denotation* (i.e. the set of its *designata*).

Fifth, I assume that my Reader accepts the view, according to which denotations of sentences are not their *nominata* (i.e. Gottlob Frege's meaning) but only parts of the latter, i.e. situations.

Sixth, and finally, I assume that my Reader accepts the view according to which the relations of connoting and denoting are purely *conventional*, whereas the relations of determining (which hold between connotations and corresponding denotations) are not.

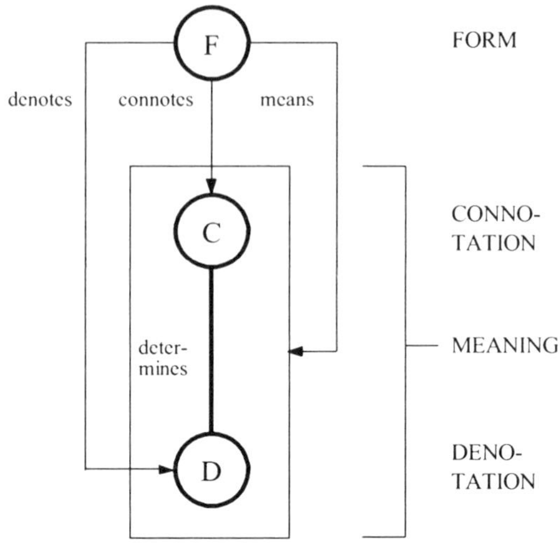

Fig. 9. Aspects of a symbol

12.2

Let us consider the intersection of two classifications of semiotic objects: existing and non-existing, on the one hand, and known (or given to us) and unknown, on the other hand. Let us remember, that semiotic objects (i.e. elements of the semiotic triad) are intentional objects. They do not exist unless they are known. Five fundamental operations can be applied to such objects: constructing, reconstructing, demonstrating, modifying, and eliminating.

Constructing is creating what does not yet exist. Let us suppose that we have one aspect of a certain symbol (e.g. its connotation) *ready*, and we create another aspect of this symbol (e.g. its form). This kind of creating — i.e. creating a certain *correlate* of a given semiotic aspect — will be called *correlating*.

Reconstructing is discovering what exists but what is not known to us. *Recorrelating* is, respectively, discovering a hidden element of an existing semiotic correlation.

Let us consider situations in which objects (including correlations) are given (or known to us). We can simply *demonstrate* them, *modify* them (to a grater or lesser extent) or *eliminate* them. In the case of correlations, eliminating can involve some members of a given correlation but not all of them.

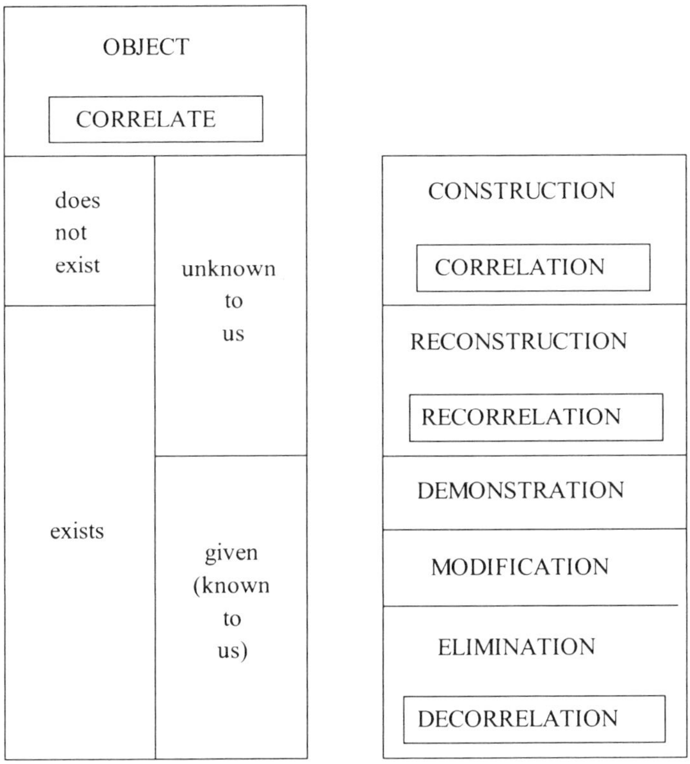

Fig. 10. Elementary semiotic operations

It can also be added that eliminating is a *maximal* modifying, while demonstrating is a *minimal* (null-)modifying.

Apart from elementary operations, there are also complex (i.e. mixed or multiplied) ones but we shall not consider them here.

Semiotic operations of demonstrating, modifying and eliminating with reference to (complex) *phrases* or *texts* concern either their substance (i.e. elements) or their structure. Symbols constitute the substance of phrases and phrases constitute the substance of texts. The structure is determined by intra-syntactic (for phrases) or inferential (for texts) rules.

I propose to distinguish (in accord with Twardowski's distinction mentioned above) between the *operation* of, e.g. defining (i.e. assigning a connotation to a given symbolic form), the *result* of defining (here: the verbal formula of the type: "*F* connotes *C*") and the *aim* of defining (here: *definiens C*).

We must remember that the same results or aims can be obtained by various operations.

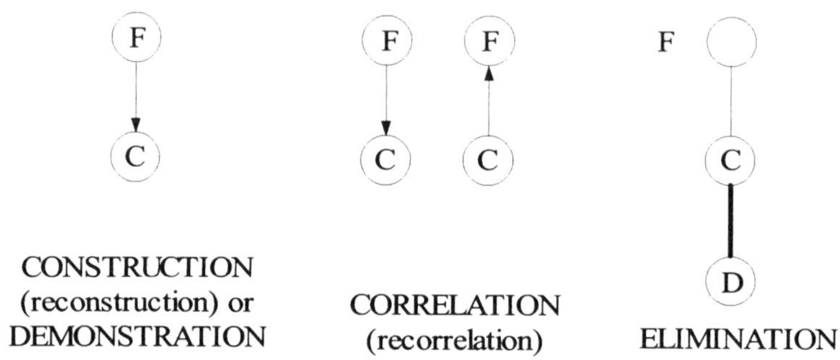

Fig. 11. Results of operations

We shall consider in detail only the operations of *constructing*, *correlating* and *modifying*.

12.3

We have seven elementary operations of this kind. Four of them, (4)—(7), create correlations.

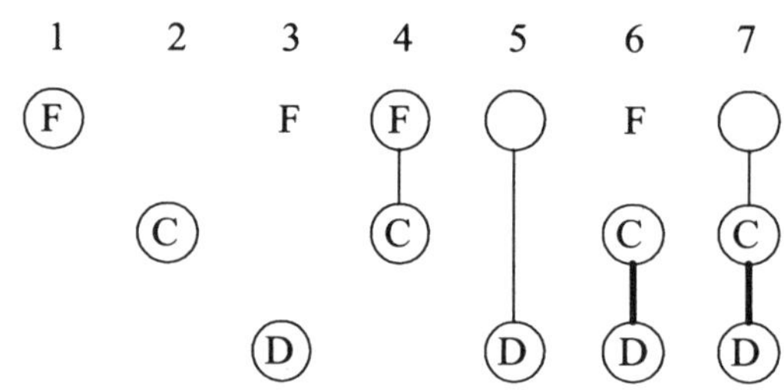

Fig. 12. Constructions

I am inclined to regard these four operations as operations of terminological POSTULATES. Ajdukiewicz's arbitrary definitions of the type:

One metre is the length of 1/10000000 of the Earth's meridian

are not, however, the verbal results of postulating (in this sense).

The only operations for which we can formulate the *criterion of correctness* (see formula under sec. 1) are (6) and (7).

12.4

There are twelve elementary operations, which consist in coordinating the given symbolic aspects with those assigned by us.

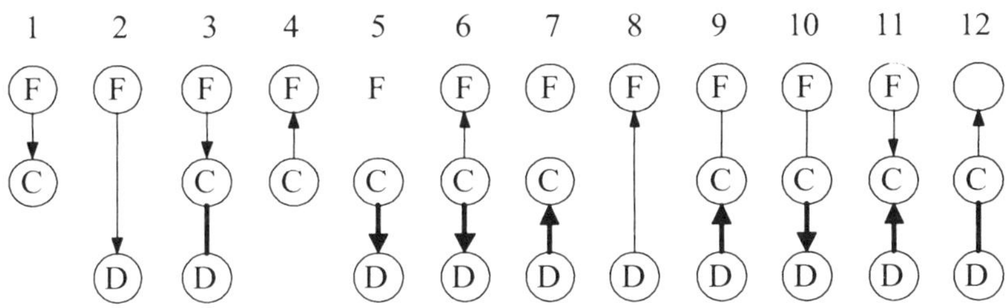

Fig. 13. Correlations (vertical coordinations)

In the case of operations (1)—(9), we have only one given aspect; in the case of (10)—(12), two aspects are given. Then, in the case of operations (1), (2), (4), (5), (7), (8), (10)—(12), we assign only one aspect; whereas in the case of operations (3), (6), (9), two aspects are assigned.

I am inclined to regard operations (1)—(3) as INTERPRETING, operations (4)—(6) as CONCRETISING, and operations (7)—(9) as CHARACTERISING. These three types of operations are distinguished in respect to the kind of given elements (respectively: forms, connotations, and denotations). On account of the kind of assorted symbolic aspects, we shall speak about TERMINOLOGISING in the case of (4), (8), (12), DEFINING in the case of (1), (7), (11), and IDENTIFYING in the case of (2), (5), (10).

The verbal expressions of particular types of correlating follow:

(1) *F* connotes C;
(2) *F* denotes *D*;
(3) *F* means *D* by *C*;
(4) *C* is connoted by *F*;
(5) *C* determines *D*;
(6) *C*, which determines *D*, is connoted by *F*;
(7) *D* is determined by *C*;
(8) *D* is denoted by *F*;
(9) *D* is determined by *C* that is connoted by *F*;
(10) *F*, which connotes *C*, denotes *D*;
(11) *F*, which denotes *D*, connotes *C*;
(12) *D*, which is determined by *C*, is denoted by *F*.

Thus, the examples from Ajdukiewicz's list of definitions are not, strictly speaking, instances of the verbal results of operations considered. His nominal definition of content:

"A polygon of nine angles" has the same content as "a hexagon"

does not indicate any content at all. It claims only that two expressions have the same content. But which one? The analogical problem arises in the case of his nominal definition of extension:

"A polygon of nine diagonals" has the same extension as "a polygon whose sum of angles adds up to 720".

Let us now consider Ajdukiewicz's examples of real definition.
(a) Definition of an individual object:

Warsaw is that town which, in 1956, is the capital of Poland.

(b) Definition of a certain kind (of objects):

Common salt is a solid with the chemical structure NaCl.

(c) Definition of a relation:

A is a brother of *B iff A* is a male and *A* and *B* have common parents. Brotherhood is the relation that holds between *A* and *B iff* etc. *ibidem*.

(d) Definition of a property:

The weight of body *A* is that property of *A*, which is common to *A* and all bodies equilibrant with *A* on a fair balance, and only to such bodies.

Classical definition:

A square is an equilateral rectangle.

Definition by postulates (implicit definition):

John was born on 29.02.1896: John was born no more then 100 years ago. John was born on Easter Sunday. John celebrated his seventh birthday on Sunday.

Inductive (recursive) definition:

The Jagiełło family is the smallest of that sets that have the following properties: (1) Ladislaus Jagiełło belongs to it; (2) if someone belongs to it, all of her/his children also belongs to it.

Definition by abstraction:

The shape of figure *A* is identical with the shape of figure *B iff A* is geometrically similar to figure *B*.

These examples can also be a base for formulating a characterisation (in our sense). Especially his implicit definition could be thus reformulated. The same goes for his descriptive nominal definition:

A square is a flat, rectilinear, equilateral, rectangular figure

Interestingly, the so-called ostensive definitions (of the type: This is so-and-so) are often used to express the result of the operation of characterising. On the other hand, the formula of (1) is meant as a verbal expression of the operation of explicating.

12.5

What we have here are fifty one elementary operations which consists in coordinating given (incomplete or complete) symbols with other (incomplete or complete) symbols.

In the case of operations (1)—(12), there is only *one* given symbolic aspect: in the case of operations (13)—(32), *two*-aspect-symbols are given; and in the case of operations (33)—(51), some symbols are coordinated with initially complete (*three*-aspect) symbols. I propose to call the operations of the first group "CONVERTING", of the second group "DECIPHERING", and of the third group "CODING".

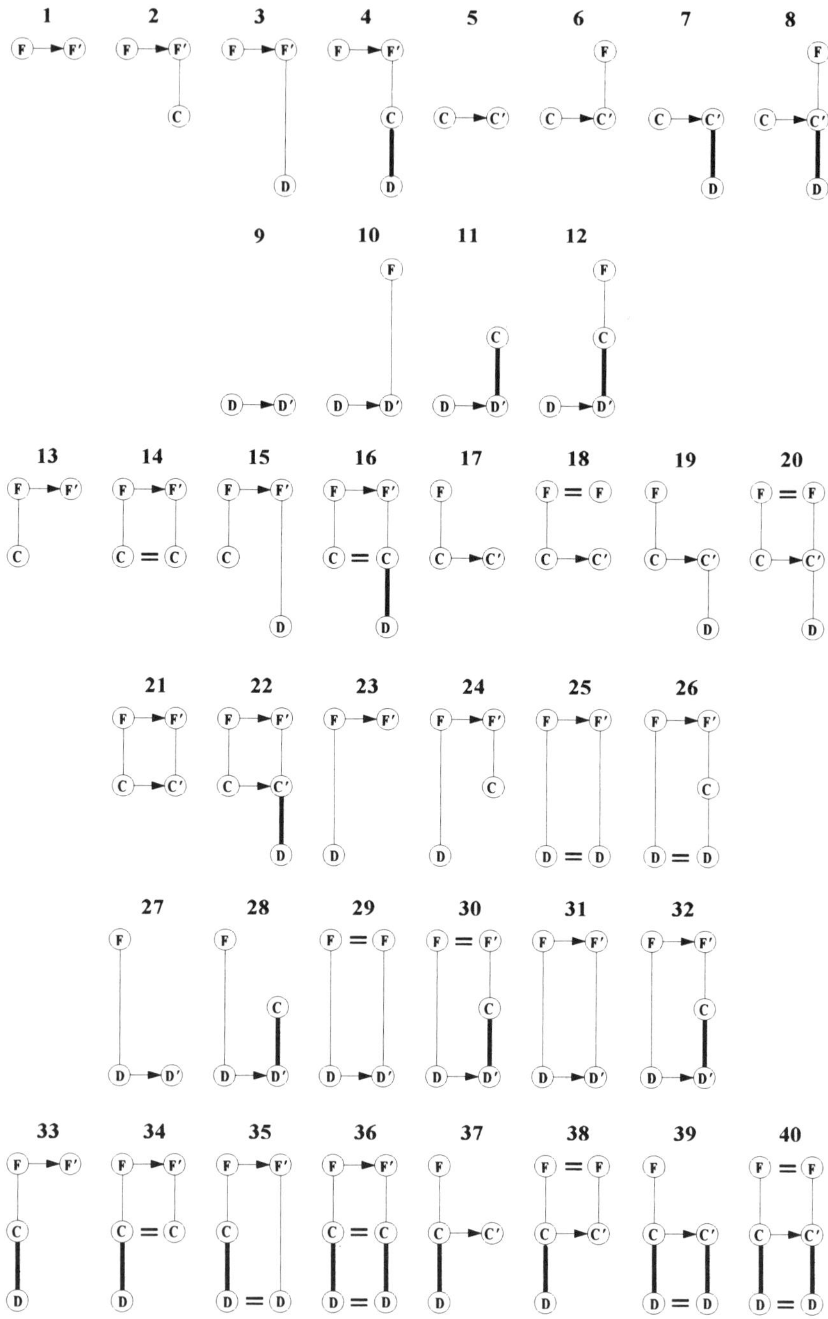

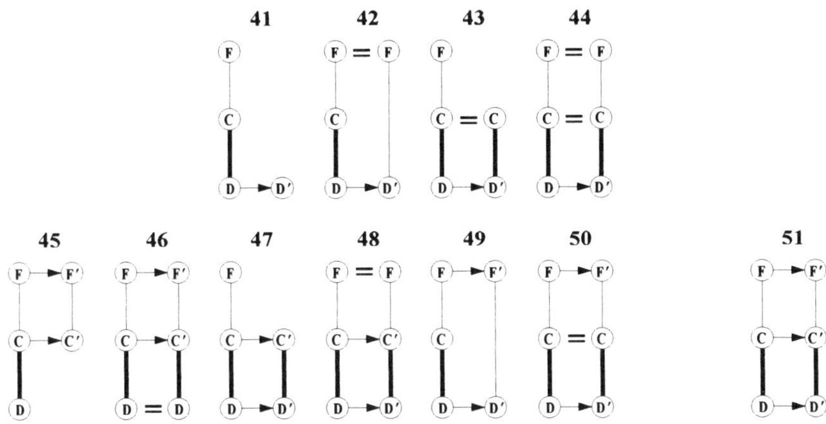

Fig. 14. Modifications (horizontal coordinations)

On the other hand, operations (1), (9), (13), (17), (23), (27), (33), (41) consist in assigning only *one* element of the semiotic triad; operations like (2), (3), (6), (7), (10), (11), (14), (15), (18), (19), (21), (24), (25), (28), (29), (31), (34), (35), (38), (39), (42), (43), (45), (47), (49) consist in assigning *two* elements; and finally, operations (4), (8), (12), (16), (20), (22), (26), (30), (32), (36), (40), (44), (46), (48), (50), (51) consist in assigning *three*-aspect-symbols. I propose to call the operations of the third group "TABELARISING", of the second group "SENSUALISING", and of the third group "SYMBOLISING".

For those operations where symbolic (or rather phrasal and textual) *forms* are modified, one can additionally perform three sub-operations:

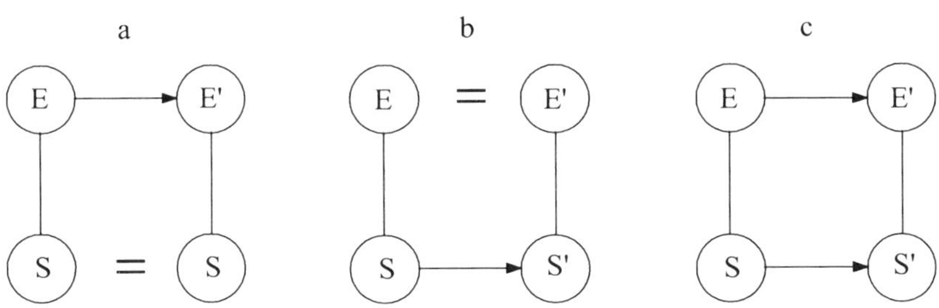

Fig. 15. Sub-operations under forms

(a) TRANSPOSITION or substantial modification (i.e. modification of formal elements);

(b) TRANSFIGURATION or structural modification (i.e. modification of formal composition);

(c) TRANSMUTATION or sub-structural modification (i.e. modification of elements as well as of the composition of phrasal or textual forms).

One can be impressed by the great plurality of possible modifications. The more striking circumstance is that this pluralism is rather unnoticed in works devoted to the analysis of the problem of definition, etc. Ajdukiewicz's works are not an exception.

When it comes to various kinds of CONVERTING, I shall only note, that it is here where we should look for situations that corresponds *de facto* to Willard van Orman Quine's radical translation as well as to Colombo's initial contacts in Cuba. On the other hand, first linguistic activities of human infants have the character of TABULARISING. In any case, I think that turning our eyes on this richness of combinations opens a prospect for solving many emerging questions.

Turning to the problem of the triad DEFINITION—EXPLICATION—PARAPHRASE, we must focus on operations of CODING (33)—(51), and especially to:

— TRANSFORMATION (36);
— TRANSCONNOTATION (40);
— TRANSDENOTATION (44);
— TRANFORCONNOTATION (46);
— TRANSCODENOTATION (48);
— TRANSFORDENOTATION (50);
— TRANSSYBMOLISATION (51).

In passing to the next problem, let us note that we can speak about the criteria of correctness only in the case of modifications that preserve at least one element of the modified object.

12.6

I propose to identify the operation of translating with transformation (36). (We read below: $D\alpha$ — denotation of α, $E\alpha$ — elements of α, $C\alpha$ — connotation of α, $S\alpha$ — composition of α.)

Translation (of *symbols*, *phrases* or *texts*), i.e. transformation

$(\alpha)(\beta)\ E\ (\alpha \text{ is translation of } \beta)\ KK\ [(C\alpha = C\beta)\ (D\alpha = D\beta)]\ A$

(a) substantial

$$E\ (E\alpha \neq E\beta)\ (S\alpha = S\beta)$$

(b) structural

$$A\ E\ (E\alpha = E\beta)\ (S\alpha \neq S\beta) \qquad (F\alpha \neq F\beta).$$

(c) sub-structural

$$E\ (E\alpha \neq E\beta)\ (S\alpha \neq S\beta)$$

Fig. 16. Translation

Word-for-word translations in particular are a substantial variant of transforming (free translations being more of a travesty). It is important to notice that Ajdukiewicz's nominal definitions in the language formulation, e.g.:

> The expression "micron" means the same as the expressions "the thousandth part of a millimetre"

(see also examples in sec. 4) are, as a matter of fact, verbal expressions (results) of translating (in our sense). On the other hand, we do not find proper definitions in ordinary dictionaries; they record usually only translational coordinations.

Besides, I would like to stress that the relation of being a translation is here irreflexive (!), symmetric and transitive.

Now, I propose to identify the operation of *explicating* with a certain variant of transforconnoting, or TRAVESTY (46), applied to (simple) symbols. It is such a travesty, where the connotation of the *explicatum* has a higher precision than the connotation of the *explicandum*.

Explication (of *symbols*), i.e. travesty + precisation

$$(\alpha)(\beta)\ E\ (\alpha \text{ is an explication of } \beta)$$

$$K\ (F\alpha \neq F\beta)\ K\ K\ (K\alpha \neq C\beta)\ (D\alpha = D\beta)$$

$$(C\alpha \text{ is more precise than } K\alpha)$$

Fig. 17. Explication

The formal characteristics of explication are the same as those of translations.

Naturally, it is an open question whether declared explications are not in fact transsymbolisations (51). It should also be added that Ajdukiewicz's regulative definitions of the type:

> The expression "salt" is synonymous (not with the expression "common salt" but) with the expression "the chemical that can be obtained from an acid by substituting a metal for hydrogen"

are not results of explications but rather of operations like (37).

It may be interesting to stress that there is a converse of explication, namely the operation of *metaphorising*.

To indicate the nature of Ajdukiewicz's paraphrases, we shall contrast them with Kotarbiński's *reductions* [Kotarbiński 1929, pp. 51ff]. The latter are structural travesties applied not to (simple) symbols (as in the case of explications) but to (complex) phrases and texts. What this means is that the elements of forms are left here (at least essentially) unchanged. The modification of forms consists here only in restructurising them.

Reduction (of *phrases*, especially sentences, or *texts*), i.e. structural travesty $(\alpha)(\beta)\ E\ (\alpha \text{ is a reduction of } \beta)$ $K\ (E\alpha \neq E\beta)\ (S\alpha \neq S\beta)\ K\ (C\alpha \neq C\beta)\ (D\alpha = D\beta)(!)$ as to categorial structure only

Fig. 18. Reduction

In Kotarbiński's reistic reductions this restructurisation tends towards the elimination of some grammatical (or semantic) categories, e.g. names of properties, relations and so on:

> "The eruption of Vesuvius destroyed Pompeii and Herculanum" is reduced to "Vesuvius erupted so that it destroyed Pompeii and Herculanum".

> "Properties exist" is reduced to "Certain objects are such-and-such".

"The relation of brotherhood is symmetrical" is reduced to "If *A* is a brother of *B*, *B* is a brother of *A*".

The decisive question of reduction in Kotarbiński's style is: Can we retain the identity of denotations after applying this operation? If not, we are again in the province of transsymbolising (51).

If I am not mistaken the relation of being a reduction is irreflexive, symmetric (!), and non-transitive.

12.7

Paraphrase, like reduction, pertains to (complex) phrases and texts. In contrast to reduction, however, this operation is a kind of substantial travesty. The difference between the connotation of *travestiens* and of *travestitum* is (as in the case of explication) a difference in the degree of precision.

Paraphrase (of *phrases*, especially sentences, or *texts*),
i.e. structural travesty

$$(\alpha)(\beta)\ E\ (\alpha \text{ is a paraphrase of } \beta)$$

$$K\ (E\alpha \neq E\beta)\ (S\alpha = S\beta)\ K\ (C\alpha \neq C\beta)\ (D\alpha = D\beta)$$

as to isomorfic structure — as to precision

Fig. 19. Paraphrase

Paraphrasing is, of course, irreflexive as well as non-transitive; but I am not quite sure whether it is symmetric or asymmetric.

What is the necessary condition for the operation of paraphrasing?

Physical phenomena	are	identical with	physical phenomena	closely connected with them
The property *A*	is	identical with	the property *B*	that has the same extension
The proposition *p*	is	identical with	the proposition *q*	equivalent to it

Our judgements	are admissible	without	any	base of	experience
From science, we	can	elimi-nate	all	elements (justifications) of	empirical character

It is admissible to draw	metaphysical conclusions	from	inquiries	about	cognition
It is admissible to draw	objective sentences (concerning things)	from	sentences	concerning	the language of the former sentences

The real	world	is exclusively	the correlate of the TRANSCENDENTAL subject, i.e. the judgements built according to transcendental norms
The true	sentences of the scientific language	are exclusively	the theses of this language, i.e. judgements built according to the deductive rules of this language

Fig. 20. Examples of paraphrases

We should have two languages. One of them is more perfect than the other:

— its vocabulary and intra-syntactic rules,
— its universe of discourse and semantic rules,
— its axioms and inferential (or deductive) rules

are *explicitly* given (in fact, they are given only partially!). The other language is not so perfect. Its corresponding components are not explicitly or even intuitively given. The appearance of non-conclusive problems as well as of paradoxes is the symptom of such an imperfection.

But what is to be done when it turns out that a certain fragment of an intended paraphrase is incompatible with (e.g. contradicts) the corresponding fragment that is in fact present in paraphrased text? Ajdukiewicz's answer is simple: the author of such a text is wrong! In my opinion, two revisions come into play here:

(1) the revision of co-ordinated (simple) symbols;

(2) the revision of the hypothesis as to the identity of syntactic, semantic, and inferential rules of the co-ordinated languages.

Thus we can never know for sure whether we speak in the same manner (i.e. by means of languages with comparable structures) about the same thing (world)!

But can we surely know to a certain extent.

Sometimes instead of a declared paraphrase, we get only a trans-symbolisation. There is no sense in looking here for any criteria of correctness. Such a semiotic modification can be evaluated only with respect to its originality (I ask my Reader not to confuse it with correspondence to the original!). We can, in fact, set up a criterion of analogy of corresponding denotations but this is a rather elusive criterion. Is the transcendental subject of idealists a good analogue to the Ajdukiewicz's set of theses (of a given language)?

On the other hand, can Ajdukiewicz be sure that his intended paraphrases are not in fact operations like deciphering (22) or even conversing (4)? Is there any difference between paraphrasing philosophical texts and fairy tales?

In conclusion, let us look at the real motivation for paraphrasing. I find two main kinds of motivation. What the people who paraphrase strive for is either to legitimise with authority or to further popularise their own scientific results.

Both of these motives have an extrascientific character.

13. ON REASONING

13.1

Let us consider the following sentences:

(1) John Field, an Irish composer, was born in Dublin.

(2) Jan Łukasiewicz, a Polish logician, died in Dublin.

(3) Field was a teacher of Maria Szymanowska, a Polish pianist, who was the mother-in-law of Adam Mickiewicz, one of the greatest Polish poets.

(4) Mickiewicz was on friendly terms with Fryderyk Chopin.

(5) In his nocturnes, Chopin referred to Field.

(6) According to Field, Chopin was *un talent de chambre de malade*.

(7) One of the greatest Polish poets married the daughter of a student of an Irish composer who was born in the town, where a certain Polish logician died, and who thought little of one of his own followers who was a friend of this poet.

Only sentence (7) is entailed by the conjunction of (1)—(6). On the other hand, it can happen that one matchs sentence (5) with sentence (6), and one accepts sentence (4) in virtue of accepting sentence (5).

Consequently, I shall use the following terminology:

(a) a reason entails its consequence;

(b) one matchs a sought sentence with a given sentence;

(c) one accepts a sentence which is not accepted as yet in virtue of a sentence already accepted (eventually in a respective degree).

Let us note that the effect of successful matching consists in accepting a sentence that states that a determinate relation (e.g. entailment, in particular) holds between a given and a sought sentence.

13.2

The starting point for the discussion about reasonings in the Lvov-Warsaw School is the classification given by the founder of the School, Kazimierz Twardowski. Using terminological conventions just introduced, Twardowski's opinion can be presented as follows:

	what is given	what is sought
INFERENCE	accepted reason	consequence
PROOF	consequence	accepted reason

Fig. 21. Kinds of reasoning according to Twardowski

REASONING = matching a reason or a consequence;

INFERENCE = matching a consequence + accepting this consequence on the base of an accepted reason;

PROOF = matching a reason + accepting a consequence on the base of an accepted reason.

13.3

The first well-developed theory of reasonings was presented by Jan Łukasiewicz. He complemented Twardowski's list of reasonings, adding to them examination and explanation, where consequences, not reasons, are given. These four reasonings — inference, proof, examination, and explanation — were divided by Łukasiewicz into deduction and reduction according to what is given: reasons or consequences.

Łukasiewicz's views can be presented in the following way:

INFERENCE = matching a consequence with an accepted sentence.

EXAMINATION = matching an accepted consequence with an unaccepted sentence.

EXPLANATION = matching a reason with an accepted sentence.

PROOF = matching an accepted reason with an unaccepted sentence.

	what is given	what is sought	
INFERENCE	accepted reason	consequence	REDUCTION
EXAMINATION	reason	accepted consequence	REDUCTION
EXPLANATION	accepted consequence	reason	DEDUCTION
PROOF	consequence	accepted reason	DEDUCTION

Fig. 22. Kinds of reasoning according to Łukasiewicz

13.4

The first significant modification of Łukasiewicz's approach was carried out by his student, Rev. Jan Salamucha. According to Salamucha, the correct classification is as follows:

	what is given	what is sought		
INFERENCE	accepted reason	consequence		DEDUCTION
EXAMINATION	reason	accepted consequence	REDUCTION	DEDUCTION
EXPLANATION	accepted consequence	reason	REDUCTION	DEDUCTION
PROOF	consequence	accepted reason		DEDUCTION

Fig. 23. Kinds of reasoning according to Salamucha

Tadeusz Kotarbiński initially accepted Łukasiewicz's theory. However he turned in the end to maintain that the «most rational» theory has a different *fundamentum divisionis* than in Łukasiewicz's theory and the characteristics of particular kinds of reasoning are a bit different. As to the classification of reasonings into deductive and reductive Kotarbiński apparently followed Salamucha's solution:

	what is given	what is sought
INFERENCE	accepted reason	consequence
EXAMINATION	reason	accepted consequence
EXPLANATION	accepted consequence	reason
PROOF	consequence	accepted reason

PROGRESSION

REDUCTION

REGRESSION

DEDUCTION

Fig. 24. Kinds of reasoning according to Kotarbiński

He replaced "deduction" and "reduction" in the Łukasiewicz's sense with "progression" and "regression", respectively. Kotarbiński supplemented Łukasiewicz's characterisation of inference and explanation with the reservation that consequence and reason, respectively, should be understood as sentences of "unknown logical value".

In the works of Tadeusz Czeżowski, there are two approaches to the issue of reasonings. The second one will be discussed later. Now, I limit myself to the first one that is based partly on Łukasiewicz's solution with the modifications by Salamucha and Kotarbiński. This classification is presented below:

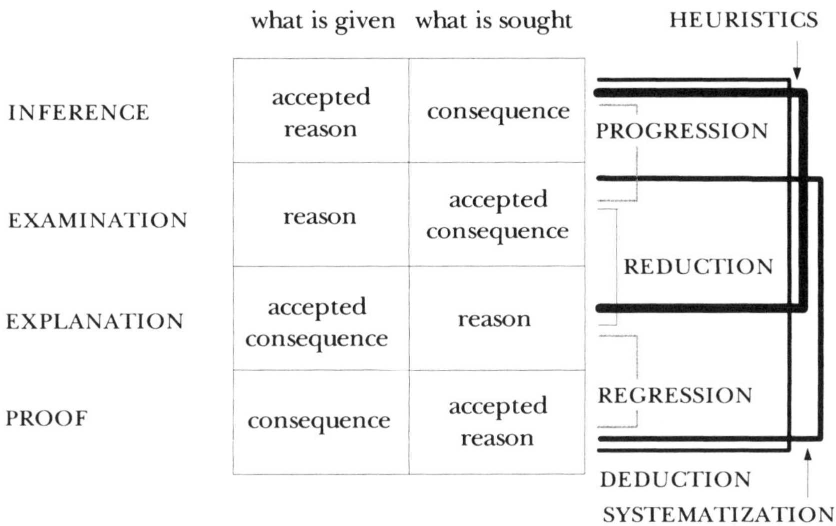

Fig. 25. Kinds of reasoning according to Czeżowski

13.5

According to Kazimierz Ajdukiewicz, the classification proposed by Łukasiewicz would not stand up to criticism.

The starting point for Ajdukiewicz's criticism was his intuition about the actual well-established meanings of terms such as "inference", "proof", "examination", "explanation", "deduction" and "reduction". He thinks that these widely used concepts are theoretically superior over the concepts proposed by Łukasiewicz. Let us illustrate this criticism regard to "inference".

First, Ajdukiewicz reproaches Łukasiewicz for the fact that, on his interpretation, one can only infer a consequence from a reason. On the ordinary understanding of the term "inference", however, not only can one also infer a reason from a consequence; one can even infer one sentence from another when there is no logical connection between them.

Second, Ajdukiewicz points out that inference defined by Łukasiewicz does not lead to the acceptance of the consequence. In other words, he claims that Łukasiewicz has not taken into account the assertive element of inference.

Third and last, Ajdukiewicz notices that Łukasiewicz's terminology restricts the extension of the notion of inference to deductive inference. As a result, all other processes commonly considered as kinds of inference (i.e. inductive inference or inference by analogy) cease to be objects of methodological study.

I do not accept Ajdukiewicz's objections.

It is true that Łukasiewicz does not say anything in his definition about accepting the consequence. However, I do not think that this is a reason for abandoning this definition. After all, it is enough to distinguish two types of Łukasiewiczian inferences: an inference where the consequence is asserted and the inference where it is not.

It is true that, according to Łukasiewicz, inferences can only be deductive. It is not true, however, that inductive reasonings have no place in Łukasiewicz's terminology at all. They constitute a kind of explanation.

13.6

Now let us make some remarks on Ajdukiewicz's positive proposal. The proposal is based on the second of Czeżowski's approaches. This second approach introduces serious changes in comparison with his first solution, mentioned above.

According to Czeżowski's second proposal, all reasonings follow the schema:

I accept "$\alpha \rightarrow \beta$".	(INFERENTIAL)
I accept "α".	PREMISES
Therefore :	
I accept "β".	CONCLUSION

Fig. 26. Schema of reasoning

Formula "$\alpha \rightarrow \beta$", which is accepted in the first premise, is a principle of reasoning. The character of this principle constitutes the foundation of the new classification, proposed by Czeżowski. According to this classification, elementary reasonings are divided into:

(1) apodeictic reasonings — in which case the principle is a theorem of classical logic;

(2) probabilistic reasonings — in which case the principle is a theorem of a probabilistic logic.

Ajdukiewicz extends the concept of reasoning to include:

(1) processes of inference;

(2) processes of derivation i.e. «non completed» quasi-inference;

(3) processes of solving tasks and problems with the use of inferences or derivations.

Ajdukiewicz's distinction between inference and derivation seems to be very valuable. However, four objections can be formulated against his general approach to inference that has been presented above.

First, the definition of inference, proposed by Ajdukiewicz, is psychologistic. The analysis of reasonings turns out to be an analysis of mental processes going on in mind of a person who infers something. By contrast, Łukasiewicz's view is radically antipsychologistic. Łukasiewicz writes as follows:

> Logic does not care [...] what mental processes are in the mind of the man who thinks logically or illogically. [...] The aim of logic is to establish objective laws that capture relations between truthfulness and falsehood of judgements [Łukasiewicz 1907, p. 64].

Second, according to Ajdukiewicz, there need not be a logical connection between the sentence that is taken to be a premise and the sentence that is taken to be a conclusion. This is tantamount not only to a psychologisation but even to a subjectivisation of the concept of inference.

Third, on Ajdukiewicz's interpretation, one could risk that any human mental activity is an inference. Thus, the denotation of Ajdukiewicz's concept of inference is broader than that of Łukasiewicz's concept of reasoning. It contains all «derivations» of some sentences from some other ones, whereas Łukasiewicz is concerned only with a scientific reasoning.

Fourth, there is a significant difficulty in evaluating Ajdukiewiczian inferences with respect to their correctness. Suppose that a person *O* accepts a sentence *p* but does not accept a sentence *q*. Suppose, further, that *O* subsequently infers *q* from *p* and so comes to accept *q* on the basis of this inference. How could we point out an error in *O*'s inference? If *O* maintains that there is a connection in *O*'s mind between *p* and *q* and that O was not sure whether p was true until O made this inference, then even if O's «reasoning» is absurd, it will still be an inference according to Ajdukiewicz.

13.7

Ajdukiewicz and Łukasiewicz have a different way of understanding not only "inference", but also "proof", "examination", "explanation", "deduction" and "reduction".

It appears that, according to Ajdukiewicz, inference is a component of all other reasonings, i.e. proof, examination and explanation. At the same time, however, as we have seen, Ajdukiewicz characterises inference psychologistically.

In my opinion, inference understood in this way cannot be a component of a scientific proof, because it allows other than logical connections between sentences or even the lack of any connection altogether. As is well known, logical proof has to be based on logical entailment.

Ajdukiewicz's definitions of examination and explanation can be criticised in a similar fashion. They rely on a concept of inference (derivation) that is not based on entailment, though it should be. These reasonings are compound and, after adequate corrections, they resemble Łukasiewicz's reasonings.

13.8

Finally, Ajdukiewicz's classification of elementary reasonings can be presented in the following way:

— deductive
REASONINGS: — probabilistic
(elementary) — logically useless

		guided by some task of		
	spontane-ous	univocal (proof) "prove that p"	bivocal (decision) "examine whether p"	multivocal (completion)
derivation	?	?	?	EXPLA-NATION
inference		PROOF *verification* *falsification*	?	?

ELEMENTS OF EXAMINATION
(complex reasoning)

Fig. 27. Kinds of reasoning according Ajdukiewicz

As is evident, the main point of Ajdukiewicz's extension is the supplementation of the list of reasonings with spontaneous reasoning as well as the inclusion of the assertive aspect of reasonings.

The distinction of spontaneous reasonings has, in my opinion, some undesirable consequences. Just like Ajdukiewicz when seeing stormy clouds can infer that rain is coming, so a dog seeing a leash could infer that it will be taken out for a walk. The only difference between these inferences is that Ajdukiewicz could express his conclusion words but the dog only with its extra-linguistic behaviour. However, I do not regard this as a fundamental difference.

If we decide that reasoning requires an active contribution of consciousness and, as Łukasiewicz put it, some kind of effort, then we should not regard spontaneous inferences to be reasonings. If we decide, however, that in spite of this objection spontaneous inferences constitute reasonings, we will not be allowed to deny that at least some of animals are capable of reasoning.

13.9

In order to grasp the peculiarities of reasonings used in science (including scientific reasonings including) the notion of reasoning must be extended in the following way:

(1) Reasoning is any operation on thoughts.

Bearing in mind our objections to Ajdukiewicz's approach to the notion of inference, let us stress that I extend here the notion of «reasoning» but not the notion of inference.

Thoughts are expressed by means of linguistic expressions, sentences in particular. I am interested here only in verbalisable thoughts; thus, for the sake of convenience, I shall speak here about sentences. For that reason, I shall give definition (1) the following form:

(2) Reasoning is any operation on expressions (sentences).

Let us consider an expression α. In the case of simple reasonings, operations mentioned in (2) can consist in:

(a) attaching any expression (e.g. β) to α;

(b) detaching any expression-member (e.g. γ) from α;

(c) dividing α into expressions-members (e.g. δ and ε);

(d) replacing any expression-member of α (e.g. ζ) with another expression (e.g. η);

(e) substituting any variable in α (e.g. ϑ) by a constant (e.g. ι);

(f) variablising — so to speak — any constant in α (e.g. κ) with variable (e.g. λ).

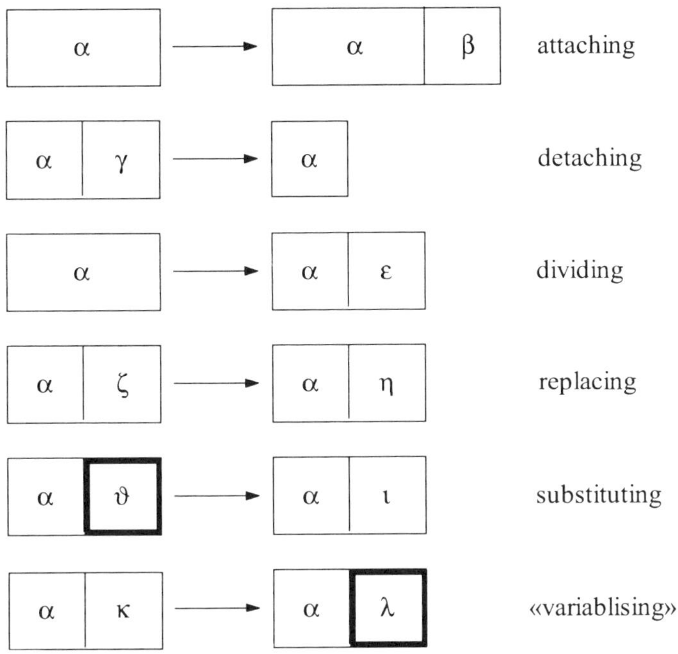

Fig. 28. Operations over expressions

Let us note that the operation of replacing can be interpreted as a sequence of the operations of detaching and attaching; the operation of substituting — as a special type of the operation of attaching; and the operation of «variablising» — as a special type of the operation of detaching.

Reasoning is sometimes understood only as the operation of attaching (a) any sentence to a given sentence. This sense of "reasoning" can be captured in the following way:

(3) Reasoning consists in deriving sentences from other sentences.

Such reasoning can be also interpreted as a kind of substitution. The schema of the resulting formula would look like this: "*Rpq*", where "*R*" is a sentential functor of the category *s/ss*.

It is easy to notice that, in the case of such formulae, there exist the following precisely determined combinations of variables and constants:

(*)	*p*	*R*	*q*
(a)	?	?	*q*
(b)	?	*R*	*q*
(c)	?	*R*	?
(d)	*p*	?	*q*
(e)	*p*	*R*	?
(f)	*p*	?	?
(**)	?	?	?

Fig. 29. Combinations of variables and constants

I ignore the versions indicated by (*) and (**) because nothing must be substituted in (*), whereas everything is to be substituted in (**) and such reasoning would be a kind of «free play of fancy».

As can be seen, reasoning in the sense of (3) is limited to cases (b) and (e), i.e. to those combinations, where the role of a constant is played by a functor and one of its arguments, whereas the second argument plays the role of a variable. It is limited further only to the cases with the functor of entailment and sometimes only to the cases with the functor of logical entailment.

I would prefer, however, to recognise as reasonings at least also:

— those operations that transform formulae (c) and (d) into sentences;

— operations on sentences that are connected by means of relations other than logical entailment.

The reason is that reasonings of this kind are *de facto* carried out even in science. N.B. Admitting the relation of equivalence allows one to consider normal definitions as reasonings.

13.10

A normal (denotational) definition of the expression *E* can be considered as consisting of two expressions *E*' and *E*" such that the denotation of expression *E* is the same as the result of a certain operation on the denotations of *E*' and *E*". In particular, it can be a sum of the denotation of *E*' and the denotation of *E*".

Thus, a classification of denotation of *E* can be considered as a definition of such a type. Thus, it is not by accident that, in practice of some sciences, the denotations of certain expressions are fixed by pointing to their subsets, which are determined by properties that constitute the *fundamentum divisionis*.

It is clear that the resulting characterisation of reasoning, for example, will be the richer, the greater the number of available classifications and thus the greater the number of appropriate properties that constitute the *fundamentum divisionis*.

It is worth noting that all views that consider only four reasonings, i.e. inference (I), proof (P), examination (Em) and explanation (Ep), notorously neglect to take into account certain possible distinctions. Usually three are mentioned:

(i) {{I, Em}, {Ep, P}} — progression/regression;
(ii) {{I, Ep}, {Em, P}} — heuristics/systematization;
(iii) {{I, P}, {Em, Ep}} — deduction/reduction.

However, to obtain the complete set of classifications for {I, P, Em, Ep}, it is necessary to consider some further classifications:

(iv) {{I}, {Em}, {Ep}, {P}};
(v) {{I}, {Em, Ep, P}};
(vi) {{Em}, {I, Ep, P}};
(vii) {{Ep}, {I, Em, P}};
(viii) {{P}, {I, Em, Ep}}.

The question arises whether there exist such properties which could constitute *fundamentum divisionis* in combinations (v)-(viii). Perhaps, in case (vii), it could be the definiteness/indefiniteness of reasons.

13.11

I would like to distinguish a classification of reasonings from evaluation of their correctness.

One can use the following criteria to assess the correctness of reasonings:

(a) truth/falsehood (i.e. truth-value) of initial (and derived?) sentences;

(b) groundedness/groundlessness of accepting initial sentences;

(c) validity/invalidity of accepting the derived sentences on the ground of the initial sentences.

If reasoning is a «science-genic» operation, which I belief, then criteria of correct reasoning can be considered to be the as negative criteria of what i scientific: what does not fulfil them falls outside the scope of science.

While I use the notion of reasoning in science in the broadest sense, I am inclined to use the notion of scientific reasoning very restrictively.

Łukasiewicz wrote in his famous work "Creative elements in science": "Not all true judgements are scientific truths" [Łukasiewicz 1912, p. 1]. I say: Not all reasonings carried out in science are scientific. And to say the truth, real creativity is connected rather with these remaining unscientific reasonings.

14. ON SIMPLICITY

14.1

Władysław Tatarkiewicz, the greatest Polish historian of philosophy, said:

> One of the intentions of my writing was to simplify the picture of the world [...] and through this simplification — to make the world clearer and more intelligible [Tatarkiewiczowie 1979, p. 157].

Some philosophers attribute more important functions to simplicity:

(a) they consider ontological simplicity as a necessary condition of doing science;

(b) methodological simplicity (of a theory) is, in their opinion, a criterion of choice between verisimilar theories.

It seems worthwhile to analyse the notion of simplicity itself.

14.2

The starting point of this analysis is the definition of *simplicity* given in Witold Strawiński's monograph *Prostota, redukcja, jedność nauki* [Simplicity, Reduction, Unity of Science] (cf. [Strawiński 1991]). Strawiński uses three specific predicates:

(a) ... is identical with ... (*s/nn*),

(b) ... is a (proper) part of ... (*s/nn*),

(c) ... is a composition of ... and ... (*s/nnn*).

To simplify (*sic*!) a little, his definitions of simplicity, complexity and the relation of being more simple than are as follows:

(1) *x* is simple *iff x* has no parts.

(2) *x* is complex *iff x* has parts.

(3) x is more simple than y *iff* y is a composition of x and z.

This last definition can be simplified (*sic!*) to the form:

(4) x is more simple than y *iff* x is a proper part of y.

Definition (3) is beset, in my opinion, with the following difficulties:
(a) it refers to an enigmatic operation of composition;
(b) it gives no possibility of:
(i) comparising objects built out of non-identical (simple) parts;
(ii) comparising objects with respect to their non-mereological parts
(e.g. properties).

Let us first consider the following situation:

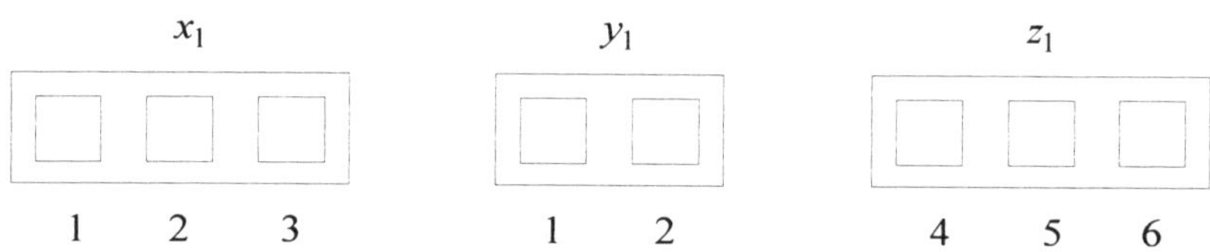

Fig. 30. Objects built out of non-identical (simple) parts

In the light of the definition under consideration, object y_1 is more simple than x_1 because y_1 differs from x_1 only in that it lacks of the mereological part 3. But what about the relation between y_1 and z_1? These two objects consist of two similar but numerically different mereological parts.

Now, let us consider the following situation:

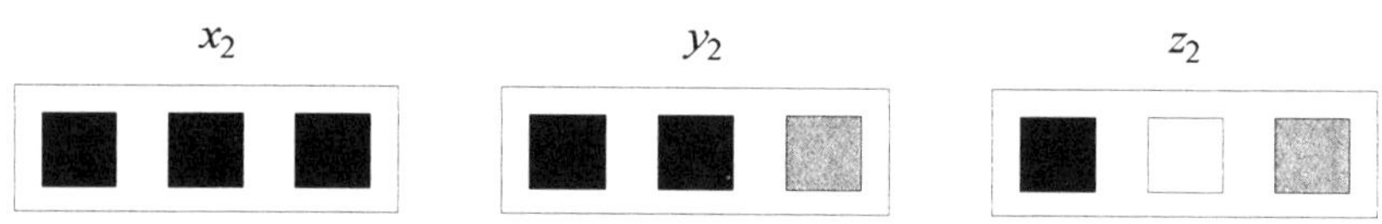

Fig. 31. Objects varying with respect to their non-mereological parts (e.g. properties)

The definition under consideration does not allow one to express the fact that x_2 is more simple than y_2 or the fact that y_2 is more simple than z_2. All these objects consist of three mereological parts. However, in the case of x_2, its three parts have the same property (of being black); in the case of y_2, only two of its parts have a common property (one part is grey); in the case of z_2, each of its parts is different with respect to its properties (the first part is black, the second one is white and the third one is grey).

14.3

My own definition of simplicity refers to one specific predicate, which needs a comment: "... is a part of ..." (*s/nn*).

The word "part" is meant here so generally that the proposed definition will not decide whether it concerns «partitive»[1] parts (i.e. components of objects) or attributive parts (i.e. properties of objects). Thus, I extend the widely known practice (cf. e.g. [Borkowski 1980, p. 11]) of interpreting the constant "$\in$" as "belongs to".

I also assume that:

A component of the object *O* = a (possibly proper) part of the object *O* of an identical ontic category as that of the object *O*.

According to this assumption, e.g. a «piece» of a given thing is a component of this thing, a subset of a given set is also a component of this set, but a property of a given object is not a component of this object. A similar notion was used by Aristotle, when he talked about ultimate components of things, i.e. about the elements [Aristotle –323, pp. 90—91].

Thus, only linguistic expressions can be components of linguistic expressions. This condition can be sharpened — the identity of semantic category can be required. Then, for example, only a sentence could be a component of a given compound sentence; neither functor nor names could play the role of components of such a sentence.

It is an open question whether physical particles are components of things because it is not clear what their ontic category is (or perhaps because the notion of an ontic category itself is vague).

It is of utmost importance to be aware of the precise sense of the terms used in this context. This is made evident by the trap into which John Duns Scotus fell when he tried to justify the (theological) thesis

[1] The term "partitive" is a derivative of the term "partition". I do not use the term "mereological" because I do not want to decide whether the parts in question are identical with the parts characterised by Leśniewski's mereology.

that God is a simple (*simplex*), i.e. a non-complex being. *Doctor subtilissimus* argued in the following manner [Duns Scotus 1299, pp. 76—78]:

God is a simple being,
because:
(i) God is (in all respects) infinitely perfect,
and thus:
(ii) every conceivable part would be infinite;
whereas:
(iii) no infinite object can be a part of something,
because:
(iv) "a part is smaller than the whole" ("*pars minor toto*").

The key issue in this argument is, of course, the logical value of premise (iv).

14.4

The predicate "... is a part of ..." and, more precisely, the related predicate "... has parts ... and ..."[2] has two meanings.

On one («synthetic») meaning:

x has parts *y* and *z* *iff* *x* has been built out of *y* and *z*.

Parts in this case are, in principle, separable.[3] This notion of part was used by St. Thomas Aquinas when he talked about the elements [Aquinas 1265, p. 85].

On the other («analytic») meaning:

x has parts *y* and *z* *iff* *x* can be divided into *y* and *z*.

Such an explication of the «analytical» meaning settles nothing about to the manner of the (potential) division. In particular, it is not decided here whether this division has a physical or an intellectual character. In the latter case, parts can be (physically) inseparable.

---

[2] The schema is not meant to settle that this predicate has *precisely* three arguments; the predicate in question has *at least* three arguments.

[3] For simplicity (*sic*!), I ignore here the problem of the status of mixtures of liquids or gases.

The fact that these two concepts of having parts are essentially different is exemplified by the fact that *this* sheet of paper can be divided (torn up) into different pieces but it has not been build from these pieces.[4]

14.5

I propose to define simplicity, complexity and the relation of being more simple than in the following manner:

(5) x is SIMPLE with respect to V *iff* x has no parts of kind V.

(Henceforth, I will place the index "V" near related expressions instead of the phrases "in respect of V" and "of a kind V".[5] I should stress that I deal with proper parts only.)

(6) x is COMPLEX$_V$, if x has parts$_V$.

I assume now that:

(7) Every simple object is more simple$_V$ than any complex$_V$ object.

(8) If x and y are complex then:
x is [n times] MORE SIMPLE$_V$ than y *iff* the number of (simple) parts$_V$ of x is [n times] smaller than y.

To distinguish simplicity/complexity characterised in (5)—(8) from simplicity/complexity discussed below, I will call the former "global simplicity/complexity". Simplicity in Strawiński's sense is a special case of such global simplicity. Recall that, according to Strawiński's definition (3), x is more simple than y *iff* y is a composition of x and z. Thus, if

y has n parts, x has m parts and z has k parts,

then

$n = (m + k)$ and $k \neq 0$,

and thus

$m < n$.

Objects that are simple/complex with respect to the multiplicity of components will be henceforth called "partitively simple/complex".

---

[4] Again for simplicity (*sic!*), I ignore the question of physical reversibility of the respective processes. This question is clearly pertinent in this context.

[5] At least some kinds can be considered to be ontic categories.

Objects that are simple/complex with respect to the multiplicity of properties will be called "attributively simple/complex". I will also speak about synthetic or analytic simplicity/complexity, respectively.

Of course, one can consider global simplicity/complexity of various objects: not only of things but also of properties and relations, for example. After all, properties as well as relations sometimes have components and properties.

It seems to me that one cannot find an equal measure/weight of complexity for *different* respects. The reason is that our intuitions in general fail in such cases.

On the other hand, we can say:

(9) *x* is more simple *simpliciter* than *y iff x* is more simple than *y* in every respect.

14.6

The question about simplicity/complexity can be repeated with reference to:

(i) particular (partitive as well as attributive) parts of a given object, i.a. a shape of an object;

(ii) structure of such an object — its composition (i.e. net of the connections among its parts) and its equipment (i.e. the matter of these parts).

In all these cases, the relevant notions can be easily rendered in terms of the notion of simplicity defined above. In order to simplify (*sic*!) further considerations, it is worthwhile to give an explicit definition of simplicity/complexity of shape (for plane figures) as well as of composition and equipment.

(10) If *x* has a MEAGRE SHAPE then the line that determines *x*'s shape has no «twists» (i.e. no changes of component directions).[6]

(11) If the line that determine shape of *x* has «twists», then *x* has a RICH SHAPE.

[6] Of course, the notion of a twist should be made precise. There are well-developed geometrical theories which explicate common intuitions connected with this notion. (N.B. These common intuitions go better with polygons than with curves, for example.)

(12) If x and y are complex with respect to their shapes then the shape of x is more meagre than the shape of y *iff* the line the determines the shape of x has more «twists» than the line that determines the shape of y.

Henceforth, objects of meagre/rich shape will be called "figurally simple/complex". Thus, the distance from A to b is the simpler figurally, the smaller the number of turns (which does not mean, however, that it is thereby the shorter). N.B. For (regular) solids, the number of tops and edges is taken here into account, which results, of course, in differing evaluations of their figural complexity [Strawiński 1991, p. 105].

Accordingly:

(13) If x is complex$_V$, then x has a LOOSE$_W$ COMPOSITION *iff* there are no relations$_W$ between parts$_V$ of x.

(14) If x is complex$_V$, then x is of a COMPACT$_W$ COMPOSITION *iff* there are relations$_W$ between parts$_V$ of x.

(15) If x and y are complex$_V$ and compactt*w*, then the composition of x is more loose$_W$ than the composition of y *iff* the number of relations$_W$ between parts of x is smaller than the number of relations$_W$ between parts of y.

Let us consider the following situation (the symbol "⇒" signifies the relation of reacting):

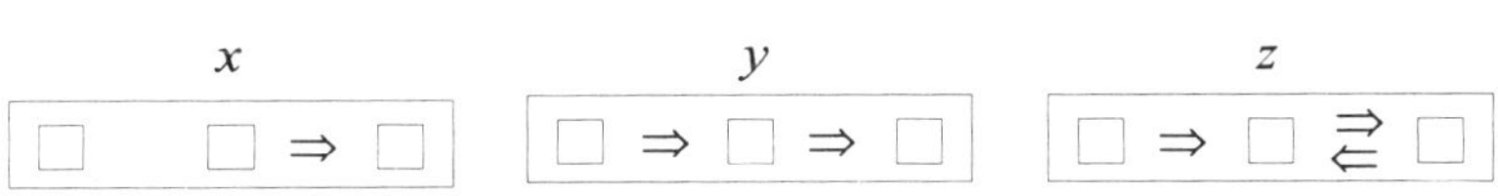

Fig. 32. Objects varying with respect to reacting

Object x has a looser composition with respect to reacting than object y, and object y has a looser composition in this respect than object z.

Henceforth, objects of loose/coherent composition will be called "structurally simple/complex". Structurally complex objects constitute SYSTEMS. The condition of structural simplicity/complexity of a given object is, of course, its global simplicity/complexity.

It is structural complexity that Aristotle had in mind when he wrote about "unity" [Aristotle –323, p. 95].

One could think about a measure of the structural complexity of a given object, which could be given by the sum of the results of multiplying the number of its internal relations of particular kinds by the measure of the complexity of a given kind. Let us notice that one of the differences in logical notations concerns just the degree of structural complexity; *cf.* the expression "$(p \wedge q) \rightarrow r$" and its analogue in Polish notation — "*C K p q r*".

Now:

(16) If *x* is complex$_V$, then *x* has a HOMOGENEOUS$_W$ EQUIPMENT *iff* all the parts$_V$ of *x* are alike$_W$.

(17) If *x* is complex$_V$, then *x* has a HETEROGENEOUS$_W$ EQUIPMENT *iff* not all of *x*'s parts$_V$ are alike$_W$.

This last definition can be simplified (*sic*!) to the form:

(18) If *x* is complex$_V$, then *x* has a HETEROGENEOUS$_W$ EQUIPMENT *iff x* has parts$_V$ of various$_W$ kinds.

Finally we have:

(19) If *x* and *y* are complex$_V$ and they are of various$_W$ equipment, then the equipment of *x* is more homogeneous$_W$ than the equipment of *y iff x* has a smaller number of various$_W$ kinds of parts$_V$ than *y* does.

Henceforth, objects with homogeneous/heterogeneous equipment will be called "materially simple/complex".[7] A materially simple object constitutes an AGGREGATE.

A measure of the material complexity of a given object could be, for instance, a sum of results of multiplying the number of components with particular kind of properties by the measure of the complexity of the properties of a given variant.

---

[7] The term "a materially simple/complex object" has some faults here and in other contexts. The problem is that, e.g., abstracts can also be considered with respects to material simplicity/complexity, whereas there are terminologies, in terms of which at least some of those objects are deemed nonmaterial. Thus the term "a qualitatively simple/complex object" might be more appropriate; but this term is flawed in other respects (some properties are not qualities, according to some people).

Let us consider the following situation (the symbol "⇒" signifies the relation of reacting, whereas the relation of «gluing» is signified by the symbol "⇓"):

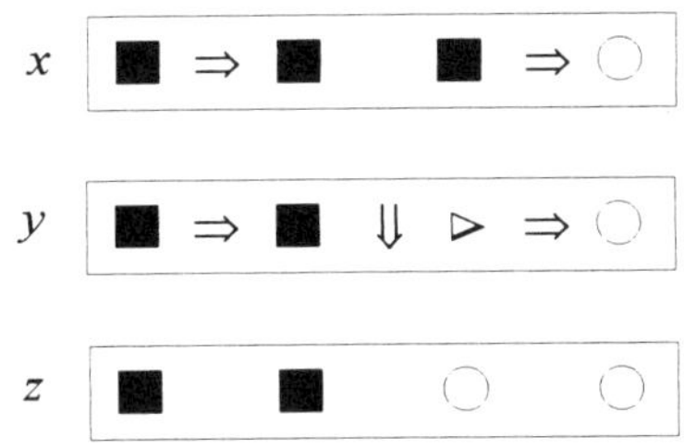

Fig. 33. Objects varying in many respects

Object *x* is here materially more simple with respect to the shape of its components than *y*, but *x* is structurally more complex with respect to internal relations (i.e. with respect to relations between its parts) than *y*. By the way, the problem arises how to formulate a criterion that would allow one to recognise object *x* as materially simpler with respect to the shape of its components than object *z*, which is — from another point of view — structurally simple.

The history of physics shows that physicists have a greater predilection for material simplicity than for partitive or structural simplicities: they would prefer it if the world consisted of a great number of objects, connected by a great number of relations, but were as homogeneous as possible.

With the aid of definitions (5)—(19), I am ready to interpret all the (known) contexts, in which the notion of simplicity occurs. This concerns also contexts situated so far from our consideration as ... the simplicity of a person: simpleton (*scil.* having only a limited education), ordinary (*scil.* of an only limited refinement), as well as ... a simple-minded one (*scil.* having only limited «gumption»).

In what follows, we will discuss two philosophically essential «dimensions» of simplicity: the ontic and the semiotic one.

14.7

Let us note at the beginning that if there is only one world then it *is* either simple or complex in a given respect. Only if there were at least two worlds, could we establish which of them is simpler in a given respect than the other one.

The question of simplicity/complexity of the world, i.e. of ontic simplicity/complexity, appears i.a. in two important contexts: (i) the context of existence of regularities and (ii) the context of the so-called homogeneity of time and space.

Let us understand by "regularity" the class of at least two «essentially» similar (i.e. alike) dependencies. World U_1 with regularities would be structurally more simple than world U_2, which contains only «unrepeatable» dependencies or is a structurally simple chaos. (By the way, the existence of regularities in the world makes it possible to present a syntactically simpler description of this world; cf. below.) From another point of view, world U_1 would be at the same time categorially more complex than world U_2 (apart form the ontic category of events there would also be the ontic category of regularities in it).

If the notion of homogeneity of time and space is to be understood in terms of the fact that regularities are *always* and *everywhere* the same, then the existence of such a homogeneity would diminish the structural complexity of the world in a greater degree. N.B. This situation could also be described in such a manner that we could talk about the material simplicity of time and space with respect to regularities: the same regularities would occur in every temporal and spatial section of the world.

14.8

The question of the simplicity/complexity of a *description* of the world, i.e. the question of the semiotic simplicity/complexity, can be reduced to the question of the simplicity/complexity of the *language* of this description and the *theory* expressed in this language. It is worthwhile to analyse this question on three levels: syntactic, semantic and pragmatic.

To determine the degree of syntactic complexity of a given language, we should take into consideration the complexity of:

(i) its vocabulary, namely:

(a) the number of (logical) constants, predicates and variables (or quantifiers), i.e. the partitive complexity of this vocabulary;

(b) variety of properties, especially formal properties (i.e. reflexivity, symmetry, transitivity etc.) of predicates, i.e. the material complexity of this vocabulary;[8]

(ii) its rules of formation, i.e. the structural complexity of utterances.

Our intuitions concerning the attributive complexity of predicates tend to vary (for example, Adolf Lindenbaum considers transitivity as an indication of the simplicity of a given predicate, whereas Nelson Goodman's opinion is just the opposite (cf. [Strawiński 1991, pp. 61ff]).

In determining the syntactic complexity of the theory itself, we take into consideration the number and the complexity of:

(i) the theorems of this theory, including axioms, definitions and hypotheses, i.e. the partitive and the material complexity of the theory;

(ii) the rules of inference accepted in the theory, i.e. its structural complexity.

Let us note, incidentally, that if "a more simple theory" in Karl Popper's sense were to mean the same as "a theory that indicate a connection previously not identified" [Grobler 1993, p. 148] then, from my point of view, such a theory has a greater syntactic partitive complexity.

In order to compare the theorems of a certain theory with respect to their partitive and material complexity, one must translate these theorems so that they containe i.a. the same logical constants or a certain set of quantifiers. Such a translation could consist in the reconstruction of respective formulas either in normal forms or in forms that contain (from among the connective constants of propositional calculus) only conjunction and negation; that last form is desirable, i.a. because of the «simple» ontological interpretation of these constants. For instance, a measure of the syntactic complexity of a given formula F could be either the length of corresponding formula in a normal form or the degree of complexity of the simplest formula equivalent to formula F. If one accept such a solution, it is easy to explain, e.g. why an independent set of axioms is commonly considered to be more simple or more «thrifty» [Wójcicki 1982: 216] than a dependent one.

---

[8] See footnote 7, above.

Consider a system that consists of two axioms "α" and "β". If axiom "β" can be derived from axiom "α" then the system does not in fact have the form:

$$\alpha \wedge \beta,$$

but of the form:

$$(\alpha \wedge \beta) \wedge (\alpha \rightarrow \beta),$$

or:

$$(\alpha \wedge \beta) \wedge \sim (\alpha \wedge \sim \beta),$$

which is, of course, syntactically more complex than:

$$\alpha \wedge \beta.$$

The problem of the semantic simplicity/complexity of a theorem/theory is connected with the scope (or the object, or better: the objective domain) of this theorem/theory. Thus:

Theorem/theory T_1 is semantically more simple than theorem/theory T_2 *iff* the object of theorem/theory T_1 is more simple than the object of theorem/theory T_2.

From a methodological point of view, it is most interesting to compare in this respect such theorems/theories whose objects stand in the relation of being a part to each other. Theorem/theory T_1 whose object has the object of theorem/theory T_2 as a part is sometimes considered to be a generalisation of theorem/theory T_2.

For that reason, a generalisation of a theorem/theory results in an increase of its semantic partitive complexity.

In particular, the greater (the more numerous) is the scope of explanation (excluding hypotheses *ad hoc*), the more general is a given explanatory hypothesis. Thus, a hypothesis of the form:

$$(x)\ C\ (Px)\ (Sx),$$

for example, the hypothesis:

"Every woman lies"

is semantically simpler than a hypothesis of the form:

$$(x)\ C\ (Rx)\ (Sx),$$

where $R = P \cup Q$, and $Q \not\subset P$, i.e. more simple than e.g. the hypothesis:

"Everybody lies".

It should be remembered, however, that some philosophers regard as more general that theorem/theory which implies another. In the language of the model theory, we would say that the class of models for the more general theorem/theory is included in the class of models for the less general one. On such an approach, generalisation of theorems/theories is accompanied by an increase in their semantic simplicity.

When we talk about the pragmatic simplicity/complexity of theorems/theories, we have in mind an assessment, first of all, with respect to:

(a) their intelligibility,

(b) ease of remembering them,

(c) ease of determining their truth-values.

The factors that influence this kind of simplicity/complexity are:

(i) the degree of the syntactic and the semantic complexity of a given theorem/theory,

(ii) the degree of the complexity of actions (with respect to the number of their components and the extent of the effort), that leads to understanding, remembering or testing a given theorem/theory,

but also

(iii) dispositions of the person who understands, remembers, or tests.

The problem of pragmatic simplicity/complexity is in fact a special case of the problem of practical simplicity/complexity, which appears in the course of solving a certain task. Here also applies the rule: do as much as possible with the aid of as little as possible.

14.9

As Popper reminds us:

> The followers of Mach, Kirchoff, and Avenarius have tried to replace the idea of a causal explanation by that of the "simplest description". [...] As it is supposed [in this conception] to explain why we prefer a description of the world with the help of theories to one with the help of singular statements, it seems to presuppose that theories are simpler than [the conjunctions of; my emendation — JJJ] singular statements. [...] If [...] we assume that theories are to be used for the sake of simplicity, then, clearly, we should use the simplest theories [Popper 1934, p. 136].

It is clear that this passage concerns partitive simplicity but not, for example, material simplicity. In the case of a conjunction of particular sentences, only one «theoriogenic» copula occurs; in the case of theories *sensu stricto*, there are more such copulas.

By the way, contrary to common opinion, the *principle of economy of thinking* does not provide an alternative criterion of truthfulness but only a criterion of choosing between verisimilar theories (grasping facts by generalisation). *Simplex sigillum veri* [simplicity is the seal of truth] at most in the domain of regularities. This rule can be introduced in terms of our terminological apparatus in the following manner:

(i) given two theories concerning the same object, choose the one that is syntactically more simple,

and

(ii) given two theories with the same degree of syntactic complexity, choose the one that is a generalisation of the other and thus is also semantically more complex than the other.

14.10

Let us assume that:

Simplification of complex object *O* = removal of some parts of object *O*,

where the removal can have a physical character, i.e. can be a physical elimination of these parts, or a mental character, i.e. can be a mental abstraction from them.

As far as simplification of the world is concerned, in principle, it can concern only some fragments of the world (in particular, its temporal or spatial sections) and various respects (we can consider murder, e.g. as a mainly structural «simplification» of its victim). The same concerns the converse operation of «complication».

From a methodological point of view, what is most important is not, of course, the simplification of the world but the simplification of its *picture*.

The examples of a syntactic simplification of a theory are:

(i) the partitive simplification of its vocabulary and rules of formation, including, for example, the empiricist reduction, i.e. the limitation of predicates only to empirical *resp.* observational ones (this operation is in general accompanied by a partitive complication of theorems formulated in such a simplified language);

(ii) definition (or more precisely, replacement of *definiens* with partitively more simple *definiendum*); the so-called simplification of fractions also belongs here: 1/1 is more simple than 2/2, because the latter is in fact 1+1/1+1; and 1 is more simple than 1/2;

(iii) axiomatisation, and the partitive simplification of the set of axioms itself (for instance, successive simplifications of Stanisław Leśniewski's *ontology* made by Bolesław Sobociński);

(iv) formalisation (N.B. the fact that some people consider formalisation and axiomatisation as a kind of complication and not simplification is due to the fact that, in such cases, the main «texts» of the initial theory is supplemented with additional «texts», e.g. the conventions of the formalism the rules of inference, etc.

Let us stress once more that very rarely — or, in any case, not always — the simplification of the language (terminology) leads to the simplification of the theory (with respect to number/generality of its theorems). Let us consider the theory containing the following theorems:

(i) $(x)\ C\ (Q_1x)\ (P_1x)$:

(ii) $(x)\ C\ (Q_2x)\ (P_2x)$.

Let: $Q = Q_1 \cup Q_2$, and $P = P_1 \cup P_2$.

This does not allow, of course, for a simplification of (i) and (ii) to the form:

(iii) $(x)\ C\ (Qx)\ (Px)$.

The semantic simplification of a theory can take the following forms (among others):

(a) ontico-categorial reduction (according to Occam's razor: *entia non sunt multiplicanda praeter necessitatem*);

(b) analytical explanation (by decomposing an analysed object into more simple parts, which function in a known manner);

(c) mathematisation (i.e. presenting empirical laws in the form of mathematical interdependencies).

It is worth noticing that if the postulate of reductionism required reduction of one theory to another in the sense that the reduced theory were to be generalised, then the realisation of this postulate would not bring a simplification of this reduced theory, but a simplification of science as a whole (understood as the theory of «everything»). If the theory *T*1 were reduced to the theory *T*2 then the current state of science:

$$N = \{T_1, T_2, T_3, \ldots, T_k\}$$

is partitively simplified to the form:

$$N = \{T_2, T_3, \ldots, T_k\}$$

would take place.

Incidentally, some philosophers think that the postulate of the simplicity of a theory is rational (only) when it is assumed that the world is simple. I understand this idea as follows: since the world is not simple (i.e. is complex), the theory of this world should not — because it cannot — be simple. We should agree with such an opinion only under the condition that the theory is taken to be a complete description of reality (the theory of «everything»). But such a description cannot be made *ex definitione*. As Ryszard Wójcicki put it, "no phenomenon can be analysed «to its full extent»" [Wójcicki 1982: 38]. Thus the assumption mentioned above is useless.

A special kind of semantic simplification is idealisation, i.e. the simplification of objects of a given theory by omitting «secondary» parts of these objects (i.e. by abstracting from these parts), and by limiting the analysis to their «essential» parts. Syntactic simplicity of idealisational theorems is correlated in a certain way with the semantic idealisational simplification. Leszek Nowak describes this correlation in the following way:

> Statement t_F (which determines on what the factor F depends) is more simple than statement t'_F *iff* t_F describes in its consequent a dependence that has a smaller number of arguments («variables») than the dependence described in the antecedent of statement t'_F. [...] It follows [...] that the simpler the statement of [...] a [given] theory, the greater the number of idealising assumptions which occur in the antecedent [...]. In effect, the simpler the statement, the more «clear the shape» of the essence of a given phenomenon presented by this statement. This is because it ignores many secondary factors that disturb the effect of the main factor [Nowak 1977, pp. 100—101].

Thus, the simplification of an object does not always have to lead to a more simple description of this object.

Let the concretised theorem be of the form (i) $\ulcorner\alpha\urcorner$, and the idealisational theorem be of the form (ii) "If *W* then β", where "*W*" is an idealising condition and "β" is a certain mathematical equality. Because of the antecedent itself, theorem (ii) is longer than theorem (i). In order to assess the complexity of the two claims, it is essential to evaluate the degree of complexity of "α" and "β". If they were to have the same degree of complexity, which would be rather unusual, then (i) would be more simple that (ii).

Already René Descartes, among others, was aware of using idealisation in some sciences [Descartes 1641: 25]. Gradually, it has been realised that no scientific discipline can do without idealisation. However, this realisation is sometimes accompanied by the belief that this is a clear limitation. Marian Przełęcki writes, for example:

> It should be stated clearly that what is presented in the monograph [*scil. The logic of empirical theories*; my emendation — JJJ] is not a realistic picture, but a simplified and idealised schema of the subject under investigation. The theories considered are much too weak to be identified with actual scientific theories and our treatment of them is based on certain idealised assumptions [Przełęcki 1969, p. 4].

14.11

In his *Die zeitgenössischen Denkmethoden*, Rev. Józef Maria Bocheński mentions two "decisive factors in the establishment of laws": "the simplicity of the laws and their interconnectedness in an axiomatic system". The rule of simplicity in Bocheński is expressed in the following form:

> If several hypotheses explain a given statement, then the simplest among them should be chosen [Bocheński 1954, p. 112].

I hope that in the light of what I have said above, the fact that this rule is schematic is patently clear.

15. ON SILENCE

15.1

The expression "to be silent" can perform two semantic functions:

(i) function of the unary predicate (*s/n*), as in "*x* is silent";

(ii) function of the binary predicate (*s/nn*), as in "*x* is silent about *y*".

I consider contexts of type (i) as derivative because they can, in my opinion, be reduced to contexts of type (ii) by means of the following rule:

(1) (*x*) (*T*) *E* (*x* is silent in the period *T*) (*y*) (*x* is silent in the period *T* about *y*).

Let us notice that there is a correspondence between the predicates "*x* is silent" and "*x* is silent about *y*", on the one hand, and the predicates "*x* speaks" and "*x* speaks about *y*". However, there is no «silence» analogue of the three-place predicate "*x* speaks to *z* about *y*" (except for phrases in *licentia poetica* and special contexts of the type "*x* is silent about *y* in the presence of *z*").

I am also inclined to reduce "passing over something in silence" to "being silent about something". The following dependence seems to hold:

(2) (*x*) (*y*) *C* (*x* passes over *y* in silence) (*x* is silent about *y*).

15.2

Everyone agrees that being silent and speaking are essentially connected. There is no common agreement, however, as to what that connection consists in. To throw some light on this matter, we shall specify necessary and sufficient conditions for both these phenomena. Here and below, I am concerned with verbal speaking (and «not speaking»[1]). However, the results of my analysis (suitably modified) can be applied also to non-verbal, i.e. gesticulatory or graphical, speaking (and 'not speaking') respectively.

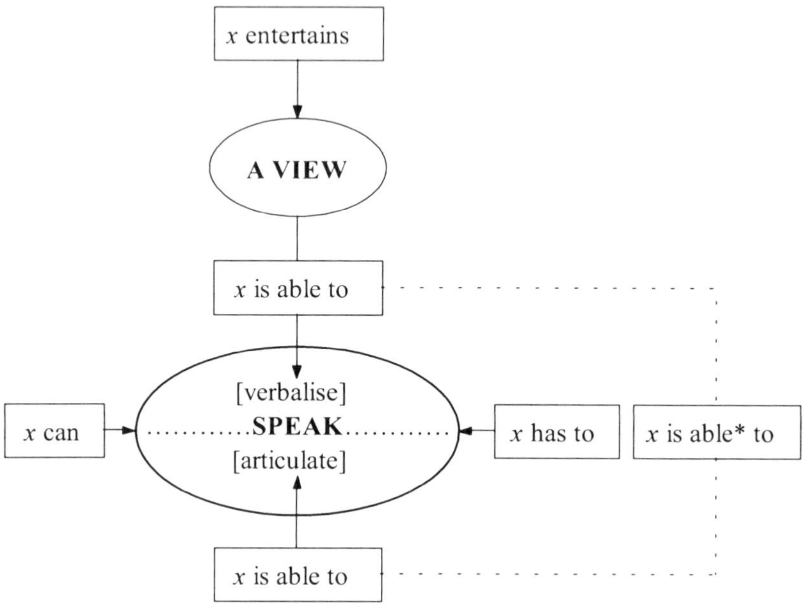

Fig. 34. The situation of speaking

I believe that the necessary and sufficient condition of speaking is the simultaneous realisation of the compulsion, the possibility, and dual ability of speaking:

(3) *E* (*x* speaks) *K* (*x* HAS TO speak) *K* (*x* CAN speak) (*x* IS ABLE TO speak).

Let us note that formula (3) is not an explicational definition of "speaking" since it contains a *circulus vitiosus*.

Being able to speak is the same as being able to VERBALISE one's views (I understand the word "views" very broadly here: it refers to beliefs and evaluations, i.e. views *sensu stricto*, as well as to «desires», e.g.

[1] I am forced to introduce this odd «negative verb» because I believe that being silent is not equivalent to not speaking (see below).

questions or commands, as we experience them), or to be able to ARTICULATE suitable speech-sounds:

(4) (x) (l) E (x is able* to speak in the language l) A (x is able to verbalise one's views in the language l) (x is able to articulate expressions of the language l).

Let us note that these abilities are mutually independent. On the one hand, one may be able to verbalise various views (i.e. in particular, put them in the form of sentences) but be unable to articulate appropriate linguistic utterances (in a given, or any, language). On the other hand, one may be able to articulate definite linguistic utterances, but not be able to verbalise definite — or, in general, any — views.

Here lies one of the sources of mysticism. It may happen that both abilities are accompanied by lack of views on a given matter. If we decide, nevertheless, to speak about it — what we produce is very often ... gibberish.

15.3

Below, by the phrase "*x* has to speak", I shall understand precisely the same as by the phrase "something pushes *x* to speak". This pushing can turn out to be ineffective in certain circumstances. ("Compulsion" so understood is semantically closer to "placing under presssure" than to "ontic necessity".) On the other hand, by the phrase "*x* can speak" I shall understand precisely the same as by the phrase "nothing keeps *x* from speaking".

Let us consider in more detail what can push *x* to speak or keep *x* from speaking.

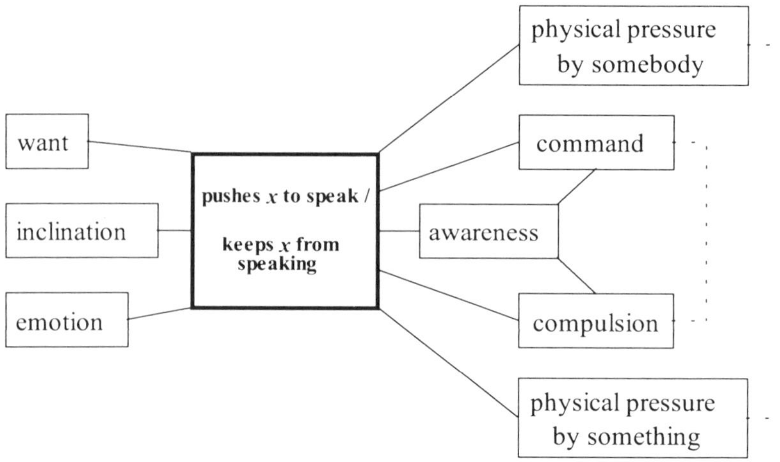

Fig. 35. Origins of speaking / «not speaking»

Thus:

(i) One can be pushed to speak (or be kept from speaking) by one's own WANT (or an act of will). In such a situation, it can be said that somebody WANTS (or wishes) to speak or WANTS not to speak. It is worth stressing that the want that pushes somebody to speak, or, on the contrary, keep her/him from speaking can spring from various motives, e.g.: pain, dreaminess, fear, hostility, slight, restiveness, vehemence, awkwardness, shame as well as their opposites.

(ii) One's own inclination can push one into a considered direction. In such a situation we say that somebody is INCLINED to speak or not to speak. Kazimierz Ajdukiewicz sharply distinguishes a situation in which somebody wants something from a situation in which somebody feels like doing something. In the former situation, he uses the term "spontaneous pro-attitude"; in the latter he uses the term "passive pro--attitude" [Ajdukiewicz 1948, pp. 204ff]. (In this paper, I try to follow to a certain extent Ajdukiewicz's beautiful analyses from the third part of the mentioned book).

(iii) A definite emotion experienced by a given person can be a direct stimulus, e.g. one of the emotions mentioned *sub* (i).

(iv) A command of another person or an «impersonal» compulsion (e.g. stemming from the norm: *Answer when asked!* — can push somebody to speak (or keep her/him from speaking, respectively). In the first situation we say that somebody is being FORCED to speak or «not speak»; in the second situation — that somebody OUGHT to speak or not to speak.

Let us note that if a command or a compulsion are to play the role of an effective stimulus, the receiver has to be aware of it. Thus, one can «not speak» either because of the lack of a suitable command or compulsion, or because of the lack of circumstances in which the compulsion arises (e.g. no question has been formulated), or because one is not aware of the appropriate circumstances (e.g. one has not noticed that a question has been asked).

(v) Finally, physical pressure applied by a person or «impersonal» circumstances can push somebody to speak (or keep her/him from speaking).

Various relations obtain between the situations descrived above as well as between their descriptions. Those relations could be presented in the form of a special calculus, which would be worth doing. Such a calculus would contain among its theorems, i.a., the following sentences:

(5) (*x*) *E* (*x* has to speak) (*x* cannot «not speak»);

(6) (*x*) *E* (*x* can speak) (*x* does not have to «not speak»);

(7) (*x*) *E* (*x* is forced to speak) (*x* is not allowed to «not speak»);

(8) (*x*) *E* (*x* is allowed to speak) (*x* is not forced to «not speak»);

(9) (*x*) *E* (*x* ought to speak) (*x* must not «not speak»).

By the way, let us notice that IT IS NOT THE CASE that:

(10*) *E* (*x* is allowed to speak) (*x* ought not to «not speak»).

We shall see at once that in the above sentences we are not allowed to substitute for the expression "not speak", the expression "to be silent".

15.4

I believe that the necessary and sufficient condition of being silent is the simultaneous realisation of the compulsion and the possibility of being silent:

(11) (*x*) *E* (*x* is silent) *K* (*x* HAS TO be silent) (*x* CAN be silent).

Let us note that formula (11) differs from formula (3) in that the necessary and sufficient conditions of being silent do not presuppose any ability. For I assume that NO ability of being silent is needed to be silent. Thanks to such an assumption, nothing stands in the way of attributing silence not only to people but also to animals and inanimate objects.

The "compulsion" and "possibility" are understood in the same way in both formulae. Contrary to Izydora Dąmbska, I think that the notion of *being silent* should not be restricted to the notion of *keeping from speaking*: the want «not to speak» is but one of the possible stimuli pushing someone into being silent. Moreover, it can happen that a given stimulus, e.g. a given emotion, pushes the same person into speaking and into being silent.

Formula (11) is analogous to formula (3) in that it IS NOT an explicational definition (in this case of "being silent") since it contains a *circulus vitiosus*.

So, what does a proper definition of "being silent" (i.e. a formula indicating the sense of this word) look like?

15.5

As I have already mentioned, it is claimed sometimes that being silent is tantamount to «not speaking»:

(12*) (*x*) *E* (*x* is silent) (*x* does not speak).

But formula (12*) is not a satisfactory reconstruction of the notion of *being silent*. Although, it is the case that:

(13) (*x*) *C* (*x* is silent) (*x* does not speak),

it is not the case that:

(14*) (*x*) *C* (*x* does not speak) (*x* is silent).

We do not speak when we sleep, but I would violate my linguistic intuitions if I said that we are silent during our sleep. We are silent only when we do not speak DESPITE the compulsion to speak.

However, it would be a mistake to try to express this intuition as follows:

(15*) (*x*) *E* (*x* is silent) *K* (*x* has to speak) (*x* does not speak).

Although, it is the case that:

(16) (*x*) *C K* (*x* has to speak) (*x* does not speak) (*x* is silent),

the converse does not hold:

(17*) (*x*) *C* (*x* is silent) *K* (*x* has to speak) (*x* does not speak),

because although the formula (13) is true, it is not the case that:

(18*) (*x*) *C* (*x* is silent) (*x* has to speak).

A certain COMPULSION is indeed a necessary condition of being silent but, according to (11), it is the compulsion to be SILENT, not the compulsion to SPEAK.

In this situation, I endorse the following analytical definition of "being silent":

(19) (*x*) *C* (*x* has to speak) *E* (*x* is silent) (*x* does not speak).

Let us note that if in formula (19) we replace the expression "*x* does not speak" with its equivalent in the form of the suitably transformed right hand side of formula (3):

(20) (*x*) *E* (*x* does not speak) *A* (*x* does not have to speak) *A* (*x* cannot speak) (*x* is not able* to speak),

then formula (18) becomes:

(21*) *C* (*x* has to speak) *E* (*x* is silent) *A* (*x* DOES NOT HAVE TO SPEAK) *A* (*x* cannot speak) (*x* is not able* to speak).

However, the expression printed in small capitals cannot occur in the formula because it contradicts the antecedent of this formula. Thus, we obtain finally:

(22) *C* (*x* has to speak) *E* (*x* is silent) *A* (*x* cannot speak) (*x* is not able* to speak).

Of course, the contradiction mentioned above arises only when both occurrences of compulsion are either of the same type or of all the indicated types. On the other hand, it is admissible that:

(23) *C* (*x* has$_1$ to speak) *E* (*x* is silent) *A* (*x* DOES NOT HAVE$_2$ TO SPEAK) *A* (*x* cannot speak) (*x* is not able* to speak).

15.6

Being silent (understood in terms of formula (19) or (22)) performs no SEMANTIC function *sensu stricto*. In particular, it states nothing: one can state something only by SAYING something, by uttering a sentence, not by merely being silent. That is why FAITH not only allows being silent but sometimes cannot manage without it. An analogous situation occurs in ART. But there is no room for being silent in the sphere of KNOWLEDGE, in science.

However, although being silent does not play any semantic function, it plays a very important PRAGMATIC function. Formulae (11) and (22) show clearly that being silent is a symptom either of the compulsion AND the possibility of being silent or of the lack of the possibility OR the lack of the ability to speak. Since that function is a source of many misunderstandings, we will take a closer look at it now.

First, let us agree that:

(24) (*x*) (*y*) (*z*) *C* (*x* is a symptom of *y* for *z*) *K* (*x* is a sufficient condition of *y*) (*x* is epistemically accessible for *z*).

If we do not identify somebody's «not speaking» as being silent, this «not speaking» cannot be a symptom of just anything. And sometimes such an identification cannot be made even indirectly especially because being silent, and passing over something in silence in particular, is very often consciously concealed.

Second, let us remember that even if one's «not speaking» is correctly identified as being silent, then it is not a symptom of either the presence or the absence of one of numerous stimuli that are relevant here, but it is a symptom of the whole DISJUNCTION of such stimuli or their lack.

These stimuli (especially the indirect ones, e.g. the motives of decisions) can not only be different but also contradictory. For instance:

(a) pleasure and pain (we speak about painful, hard, deep, mournful, depressing silence; we say that silence is a burden for somebody, that somebody sinks into or falls into silence);

(b) concentration (we speak about focused, solemn silence) and dreaminess (we speak about being inadvertently silent);

(c) bravery (we speak about heroic silence) and fear (we speak about silence caused by fear);

(d) friendship (we speak about a silent bow, handshake) and hostility (we speak about dead, stony, hostile silence; we speak about conspiracy of silence, silent indignation);

(e) appreciation (we speak about tacit appreciation or its silent signs) and contempt (we speak about ignoring something or somebody with silence);

(f) submission (we speak about enduring something in silence, tacit agreement) and restrictiveness (we speak about persistent silence);

(g) gentleness (we speak about discrete silence) and vehemence (we speak about awkward silence);

(i) pride (we speak about lofty silence) and shame (we speak about troubling, shy silence).

Only when we know in advance that a certain GROUP of disjuncts does not occur (e.g. that a silent person HAS an opportunity to speak), are we entitled to consider silence as a symptom of the occurrence of the remaining GROUP of disjuncts (in the above examples, as a symptom of the fact that this person is not able* to speak). To avoid misunderstandings, I must stress that sometimes silence is USED as a conventionalised univocal medium of expressing something, e.g. respect, and in such a case it is, of course, a symptom of this respect [Dąmbska 1971, p. 101]. Let us note that such usage is in general (if not always) preceded by a certain verbal signal (e.g. by the formula: "Let us commemorate *XY* with a moment of silence"). Here, I am not interested in such a conventionalised silence.

Thomas Carlyle wrote:

> Silence is deep as Eternity; speech is shallow as Time [Carlyle 1872, p. 25].

The «depth» of silence consists just in its semiotic indefiniteness — and to this indefiniteness the «significance» of silence can also be reduced.

CONCLUSION: SCIENCE AND CREATION

Circa 1623, Maciej Kazimierz Sarbiewski, a great Polish poet and scholar, in his treatise *De perfecta poesi*, introduced *expressis verbis* the notion of *creativity* (*creatio*) into the definition of poetry:

> A poet gives a new existence to objects which he imitates, and he apparently *creates* them again. This is because a poet, by means of some apparent *creative* art, can call into existence the subject itself as well as the mode of its presentation. [...] Other arts [...] only imitate and reconstruct because they take as a datum either their material, or their subject, or at least the means of imitation [Sarbiewski 1623, pp. 4, 6].

Quite a long time has passed before creativity has begun to be considered in Europe a definitional feature of every art.

1.

The Old Greek civilisation did not know the notion of *creativity* at all. Moreover, they did not have the notions of *science* or *art*, either. What is today called science and art, the Ancient Greeks put under the name "τέχνη", which included also trade. τέχνη was for the Ancient Greeks every human skill, i.e. the ability to do something according to stated «reasonable» rules. The language of the ancient Romans contained the word "*creator*", but it meant the same as "father". In the Middle Ages, *creatio* was the exclusive privilege of God and it meant *creatio ex nihilo* — creation from nothing. In this sense, to create something means (more or less) the same as to cause its coming into existence in a certain place and at a certain time, though it had never existed anywhere before: neither as a whole nor in any part.

Something reminiscent of this *creatio ex nihilo* remained also in the notion of *creativity* applied to art.

An artist is a creator in the sense that s/he makes something new but uses something that has existed before. A work of art is creative or new but it is not new altogether, nor in every respect. Fryderyk Chopin is a creator of the ballad *g*-minor op. 23; he composed it, say, in 1831; beforehand, it had not existed. If it were not so, the composition marked by this opus would be a mere plagiarism. But Chopin did not compose this ballad *ex nihilo* — he had some of its «parts» at his disposal: a material in the form of the «palette» of piano sounds and formal elements such as scales of the major-minor system and rules of manipulating sound material in these scales.

At the turn of the 19th century, some artists started dreaming that their vocation is to create *ex nihilo*: that the true work of art is a work that is completely new.

The dream about such complete novelty is reminiscent of the dream about constructing a *perpetuum mobile*: it is unrealisable. However, some attempts were made to realise it, e.g. works whose authors purposefully distanced themselves from everything that was «old». In particular, musicians began to create pieces for new *instrumentarium*; they refused to respect existing rules of composition and they refused to appeal to elaborated stylistic forms.

They were not satisfied when a piece was new in a certain respect only — it had to be new in all respects: unique and original.

In the end, it turned out that if every new piece were to be fully original then the notion of a rule of composition would cease to be applicable to it. If the composer her/himself determines the rules valid in her/his work and, moreover, if s/he determines them for each of her/his pieces separately (i.e. if s/he is a lawmaker for her/himself) an «external» evaluation of her/his work is impossible. One cannot tell whether s/he is a master or a bungler.

2.

Actually, until the very end of the 19th century it was a common opinion that science, in contrast to art, is devoid of any creative element. An artist creates, whereas a scientist only reconstructs: s/he discovers what there is (facts given in experience) and states the discovered facts in true sentences.

In 1912, in his paper "O twórczości w nauce" [Creative Elements in Science], Jan Łukasiewicz showed that this common opinion was mistaken.

First, not all truths belong to science. That the page on which the beginning of Chopin's ballad *g*-minor op. 23 is printed in my copy of ballads edited by Państwowe Wydawnictwo Muzyczne in 1949 is yellowed is a fact but it is not a scientific fact.

Second, not all useful and not all general truths are scientific truths. The sentence that state that I bequeathed my collection of scores (let us assume that I have indeed made such a legacy) is useful for the endowed but it does not belong to science. By analogy, a sentence that state that there is at least one quaver in each of the first three even measures of the ballad *g*-minor is general and true but it does not belong to science.

And conversely, not all scientific sentences are truths. (We will come back to the question which are not.)

Thus, what distinguishes scientific truths?

A scientific theory is a set of sentences, but a specially ordered set: the sentences that belong to it are connected by the relation of logical consequence — they constitute a system closed under this relation.

The creative element pertains to both components of a scientific theory.

First, some sentences of a theory state facts. They are uncreative in the sense that they only reconstruct what is given to the scientist in her/his experience. But some sentences are not factual: they either generalise facts or explain them in a different manner. And it is precisely the explanatory hypotheses that are a creative element in a theory: a scientist does not discover hypotheses — s/he creates them. At most s/he can consider them as well tested — but not as true *simpliciter*. (Thus, scientific hypotheses are not truths *sensu stricto*.)

Second, there is nothing in nature that corresponds to the relation of consequence, which connects sentences of a theory. Experience, in any case, is silent about such counterparts. It is thus natural to wonder, where the constructive element of a theory comes from? Since it is not discovered in nature, it must be a creative contribution of a scientist.

There is one more sense, in which one can speak about creativity in science. This sense is derivative of a certain distinction introduced at the beginning of the 20th century by Kazimierz Twardowski, the distinction between an act and an product.

Let us illustrate this distinction with a simple example. If during my lecture, I wrote something on the blackboard instead of speaking, the product of my writing would be a certain inscription. The writing (act) and the inscription (product) are not the same thing. The fact that an inscription would be on the blackboard (and that it could then be erased, e.g.), whereas the writing could not be on the blackboard (and thus it could not be erased) bears witness to this difference for example.

A scientific theory is a product, and the most important acts of which it is a product are experience and reasoning. Even if, for the sake of simplicity, we agree that experience is a reconstructive act, an act of reasoning is, undoubtedly, saturated with creativity. It is impossible, for example, to give a recipe that would tell how to find an explanatory hypothesis or a proof of a specified proposition. A hypothesis, as well as a proof has to be invented and its invention is a creative operation.

3.

Thus, it seems that the traditional belief that there is a precipice between art and science is not true — or, at any rate, the existence of a precipice is not caused by the creative character of art and the uncreative character of science. Creativity saturates equally the works of artists and the works of scientists.

One might also argue that the difference between art and science consists in the fact that an artist aspires to perfection, whereas for a scientist perfection is of secondary importance.

Perfection is a feature of human products relativised to the rules of their production: the more precisely a given product observes these rules, the more perfect it is.

The fact that there can be no art without rules — or, in any case, that the comparison of the works of art with respect to perfection becomes meaningless — was known already in antiquity.

It was not without a reason that the pieces of the Ancient Greek music were signified by the term "νόμοι" ("laws"), and in fine arts they talked about canons (measures).

Only in the 20th century did the logicians reconstruct the rule of creating scientific theories, giving at the same time criteria for evaluating their correctness and (in consequence) their perfection. The conformity with the rules has ceased to be the exclusive value of art.

It begins to look as though there is no sharp boundary between art and science. Is this really so?

4.

It is hard to answer this question to the full extent. So, let us narrow the question. Is there a sharp boundary between music and science?

Let us note that in the Middle Ages — on the verbal level — music was one of the sciences: together with logic, mathematics, astronomy, grammar and rhetoric. Moreover, the sciences have been regarded as *artes liberales*, liberated arts, i.e. arts liberated from physical effort which is necessary in the work of a sculptor or a painter. But the word "*musica*" functioned at that time as a shorthand of the expression "*scientia musicae*", i.e. it referred to the theory of music. The theory of music — or, broadly, musicology — is undoubtedly one of the sciences. But we are interested here in the music itself — and not in its science, in musical compositions — and not in their theory.

The expression "musical composition" is equivocal. It can signify: a creator's idea, a performer's interpretation and a receiver's reception, i.e. some mental objects, as well as a score and a realisation, i.e. some physical objects.

Let us consider Chopin's ballad *g*-minor op. 23. Listening to the suitable record of Witold Małcużyński, we hear a certain realisation of this ballad. Let us look at the initial fragment of one of its scores. It is a (simplified) picture of Chopin's idea. Thanks to the score, in the mind of a performer, an interpretation can be created, which reconstructs the idea of the creator and which is the basis of its physical realisation. In us, the receivers, this realisation evokes a specific reception.

We encounter a similar ambiguity in the case of the expression "scientific theory". When we say that a scientific theory is a system of sentences connected by the relation of (logical) consequence, we can have in mind either a system of sentences-inscriptions (the analogue of a score), a system of sentences-utterances (the analogue of a realisation), or a system of sentences-propositions, in particular beliefs experienced by the author (the analogue of an idea).

Below, in order not to become entangled in complex ontological problems, we shall speak about musical compositions as certain realisations, and about scientific theories as certain systems of sentences-utterances. Thus, in both cases, we will be concerned with certain physical objects.

Sentences are special linguistic expressions. At least some of the sentences that belong to a scientific theory state facts: stating is their semantic function.

In the 19th century, it has been suggested that musical compositions, too, are certain linguistic pieces. In the analyses of some musical compositions, the very notion of ...musical sentences began to appear; the term "musical language" was also introduced. Is using a language indeed something that connects musical compositions with scientific theories?

Let us do away with this myth once and for all. The so-called musical sentences are not sentences *sensu stricto*. Music performs no semantic function. Musical sentences do not state any facts: they state nothing at all. Neither do they perform the pragmatic function of expressing — in any case, not in the way language does. The sentence "Chopin's ballad op. 23 is in the mode *g*-minor" states a certain fact concerning Chopin's ballad op. 23, namely the fact that it is in the mode *g*-minor. This sentence can be used to, e.g. express the belief that this fact occurs. But neither the musical composition as a whole nor any of its fragments, including the so-called musical sentences, can be used to express beliefs.

Let us consider any «musical sentence» from the ballad *g*-minor mentioned above, for instance, the «sentence» between measures 8 and 12.

Fig. 36. F. Chopin, *Ballad* op. 23 (measures 8—12)

No fact is stated by this fragment of the ballad; no belief is expressed by the person who performs it. If music expresses anything, it expresses moods — not beliefs. And, importantly, music expresses not well-defined moods. The fragment of the ballad mentioned above expresses, in my opinion, the mood of contemplation. But if somebody contradicted me and said, for example, that it expresses regret or longing, I would not be able to convince her/him otherwise. I am not able to say precisely what is contemplation, regret or longing but I am sure that they are different: I contemplate SOMETHING, I regret that SOMEBODY did something and I long FOR something or for someone.

The analogy between a «musical language» and a language *sensu stricto* is based only on the fact that both have a distinctive «syntactic» structure. But this is not a unique feature of language: every complex object is structuralised.

5.

Perhaps, here we have found the boundary between science and music: language is a necessary component (the constructonal element) of science, but it does not essentially occur in music. Is this boundary sharp, however?

After all, there is vocal and vocal-instrumental music, where language *sensu stricto* seems to occur. In what is sang (with the use of words), it is possible to distinguish fragments in the form of sentences that perform semantic functions. But do they state any facts, as factual sentences do in scientific theories? Does a vocalist singing them express any definite beliefs, as does a scientist presenting her/his theory?

Certainly not.

If sentences that belong to vocal or vocal-instrumental compositions state anything, it is not real facts but fictional states of affairs. These sentences are, as Roman Ingarden says, only quasi-propositions; they only apparently state states of affairs. They also only appear, but do not, express beliefs. This is so in the case of vocal and vocal-instrumental compositions as well as in the case of belles-lettres.

But here a certain doubt arises.

Scientific laws, which are connected with factual truths by the relation of logical consequence, are idealisational in nature. They are simplifications of what is really the case. They hold in conditions that are not satisfied in reality.

Thus, they also contain an element of fiction.

REFERENCES

Ajdukiewicz, K. (1923). *Główne kierunki filozofii* [Main Trends of Philosophy]. Lwów: K.S. Jakubowski.

Ajdukiewicz, K. (1934a). Language and Meaning. In: [Ajdukiewicz 1977], pp. 35—66.

Ajdukiewicz, K. (1934b). The World-Picture and the Conceptual Apparatus. In: [Ajdukiewicz 1977], pp. 67—89.

Ajdukiewicz, K. (1934c). On the Applicability of Pure Logic to Philosophical Problems. In: [Ajdukiewicz 1977], pp. 90—94.

Ajdukiewicz, K. (1934d). On the Problem of Universals. In: [Ajdukiewicz 1977], pp. 95—110.

Ajdukiewicz, K. (1936a). Okres warunkowy w mowie potocznej i logistyce [Conditional in Natural Language and in Logistics]. *Ruch Filozoficzny* **14** (1—3), 134a—134b.

Ajdukiewicz, K. (1936b). Syntactic Connection. In: [Ajdukiewicz 1977], pp. 118—139.

Ajdukiewicz, K. ([1938] 1948). *Propedeutyka filozofii* [Propaedeutics of Philosophy]. Wrocław & Warszawa: Książnica & Atlas.

Ajdukiewicz, K. (1946). O tzw. neopozytywizmie [On the So-Called Neo-Positivism]. In: [Ajdukiewicz 1965a], pp. 7—28.

Ajdukiewicz, K. (1948). Z dziejów pojęcia materii [From the History of the Notion of Matter]. In: [Ajdukiewicz 1965a], pp. 61—76.

Ajdukiewicz, K. ([1949] 1973). *Problems and Theories of Philosophy*. London: Cambridge University Press.

Ajdukiewicz, K. ([1956a] 1984). On Definitions. *Dialectics and Humanism* **11** (2—3), 235—256.

Ajdukiewicz, K. (1956b). Okres warunkowy a implikacja materialna [Conditional and Material Implication]. In: [Ajdukiewicz 1965a], pp. 248—265. [Fragments as] Conditional Statement and Material Implication. In: [Ajdukiewicz 1977], pp. 222—238.

Ajdukiewicz, K. (1960). Syntactical Connections between Constituents of Declarative Sentences. In: [Ajdukiewicz 1977], pp. 269—281.

Ajdukiewicz, K. (1965a). *Język i poznanie* [Language and Knowledge], vol. 2. Warszawa: PWN.

Ajdukiewicz, K. ([1965b] 1974). *Pragmatic Logic*. Translated by O.A. Wojtasiewicz. *Synthese Library*, vol. 62. Dordrecht & Warszawa: D. Reidel Publishing Company & PWN.

Ajdukiewicz, K. (1977). *"The Scientific World-Perspective" and other Essays. 1931—1963*. Dordrecht: D. Reidel Publishing Company.

Aquinas, St Thomas ([1265] 1979). *Compendium theologiae*. In: Santi Thomae de Aquino *Opera omnia*, tomus 42, pp. 75—206. Roma: Editori di San Tommasso.

Aristotle ([-323] 1960). *Metaphysics*. Ann Arbor: Ann Arbor Paperbacks.

Aristotle ([–347] 1975). *Posterior analytics*. Oxford: Clarendon Press.

Augustynek, Z. (1975). *Natura czasu* [The Nature of Time]. Warszawa: PWN.

Augustynek, Z. (1984). Wersje materializmu [Versions of Materialism]. *Studia Filozoficzne* **27** (11—12), 3—10.

Augustynek, Z. and J.J. Jadacki (1993). *Possible Ontologies. Poznań Studies in the Philosophy of the Sciences and the Humanities*, vol. 29. Amsterdam: Rodopi.

Bacon, F. ([1597] 1955). The Essay on Truth. In: *Essays*, pp. 5—8. London: Oxford University Press.

Baier, A.C. (1967). Nonsense. In: P. Edwards *et al.* (eds.), *The Encyclopaedia of Philosophy*, vol. 5, pp. 520—522. New York: The Macmillan Company.

Berkeley, G. (1710). A Treatise Concerning the Principles of Human Knowledge. In: [Berkeley 1949], pp. 19—113.

Berkeley, G. (1714). Three Dialogues between Hylas and Philonous. In: [Berkeley 1949], pp. 147—263.

Berkeley, G. (1949). *The Works*, vol. 2. London: Thomas Nelson and Sons Ltd.

Biegański, W. (1903). *Zasady logiki ogólnej* [Principles of General Logic]. Warszawa: Kasa J. Mianowskiego.

Biegański, W. (1912). *Teoria logiki* [Theory of Logic]. Warszawa: E. Wende.

Bocheński, J.M. ([1954] 1965). *The Methods of Contemporary Thought*. Dordrecht: D. Reidel Publishing Company.

Bogusławski, A. (1986). Analiza zdań warunkowych a problem funkcji semiotycznych [Analysis of Conditional and the Problem of Semiotic Functions]. *Studia Semiotyczne* **14—15**, 215—224.

Borkowski, L. (1964). Uwagi o okresie warunkowym oraz implikacji materialnej i ścisłej [Notes on Conditional, and Material and Strict Implication]. In: *Rozprawy logiczne. Księga pamiątkowa ku czci profesora Kazimierza Ajdukiewicza* [Logical Treatises. Festschrift in Honour of Professor Kazimierz Ajdukiewicz], pp. 11—22. Warszawa: PWN.

Borkowski, L. ([1980] 1995). Pewna wersja definicji klasycznego pojęcia prawdy [A Version of the Definition of Classical Notion of Truth]. In: *Pisma o prawdzie i stanach rzeczy* [Writings on Truth and States of Affairs), pp. 10—22. Lublin: Wydawnictwo UMSC.

Borowski, M. (1921). O przedmiotach fizycznych, psychicznych, idealnych i fikcyjnych [On Physical, Psychical, Ideal and Fictitious Objects]. *Przegląd Filozoficzny* **24** (3—4), 139—163.

Carlyle, T. [1872). Sir Walter Scott. In: *Critical and Miscellaneous Essays*, vol. 6, pp. 21—81. London: Chapman and Hall.

Carnap, R. (1935). *Philosophy and Logical Syntax*. London: Kegan Paul *et al.*

Chamberlain, J. ([1906] 1972). [Quotation from a Speech]. In: *The Concise Oxford Dictionary of Quotations*, p. 59. London: Oxford University Press.

Chwedeńczuk, B. (1984). *Spór o naturę prawdy* [Controversy about the Nature of Truth]. Warszawa: PIW.

Chwistek, L. ([1930—1931] 1961). Zagadnienia kultury duchowej w Polsce [Problems of Spiritual Culture in Poland]. In: *Pisma filozoficzne i logiczne* [Philosophical and Logical Writings], vol. 1, pp. 147—277. Warszawa: PWN.

Chwistek, L. ([1935] 1949). *The Limits of Science*. New York & London: Harcourt, Brace and Company & Routledge and Kegan Paul Ltd.

Cohen, J.M. & M.J. ([1960] 1971). *The Penguin Dictionary of Quotations*. Harmondsworth: Penguin Books Ltd.

Czarnocka, M. (1986). *Kryteria istnienia w naukach przyrodniczych* [Criteria of Existence in Natural Sciences]. Wrocław: Ossolineum.

Czerwiński, Z. (1958). O paradoksie implikacji [On the Paradox of Implication]. *Studia Logica* **7**, 265—271.

Czeżowski, T. ([1938] 1959). *Główne zasady nauk filozoficznych* [Main Principles of Philosophical Sciences]. Wrocław: Ossolineum.

Czeżowski, T. (1946). Prawda a język [Truth and Language]. *Wiedza i Życie* **15** (3), 226—228.

Czeżowski, T. (1951). Identity and the Individual in Its Persistence. In: [Czeżowski 2000], pp. 218—225.

Czeżowski, T. ([1952] 1958). Uwagi o klasycznej definicji prawdy [Notes on the Classical Definition of Truth]. In: *Odczyty filozoficzne* [Philosophical Lectures), pp. 68—74. Toruń: TNW.

Czeżowski, T. (1964). On Individuals and Existence. In: [Czeżowski 2000], pp. 284—254.

Czeżowski, T. (1970). Prawda w nauce [Truth in Science]. *Studia Filozoficzne* **13** (3), 19—25.

Czeżowski, T. (2000). *Knowledge, Science, and Values. Poznań Studies in the Philosophy of the Sciences and the Humanities*, vol. 68. Amsterdam: Rodopi.

Dąmbska, I. (1931). Prawa fizyki wobec postulatu prawdziwości twierdzeń naukowych [Physical Laws in the Presence of Truthfulness of Scientific Statements]. In: *Księga pamiątkowa PTF we Lwowie* [Festschrift of the Polish Philosophical Society in Lvov), pp. 183—198. Lwów: Książnica & Atlas.

Dąmbska, I. (1971). O funkcjach semiotycznych milczenia [On Semiotic Functions of Silence]. In: *Znaki i myśli* [Signs and Thoughts], pp. 93—105. Warszawa: PWN.

Descartes, R. ([1641] 1941). *"The Meditations" and Selections from "The Principles of Philosophy"*. La Selle: Open Court Publishing Company.

Drozdek, A. (1980). Problematyka prawdy [Problems of Truth]. *Studia Filozoficzne* **23** (9), 186—189.

Drozdowicz, Z. (1980). Kartezjańskie kryterium prawdziwości [Cartesian Criterion of Truth]. *Ruch Filozoficzny* **33** (3—4), 243—248.

Duns Scotus, J. ([1299] 1966). *A Treatise of God as First Principle*. Chicago: Franciscan Herald Press.

Grice, H.P. (1975). Logic and Conversation. In: P. Cole (ed.), *Syntax and Semantics*, vol. 3 (*Speech Acts*), pp. 41—58. New York: Academic Press.

Grobler, A. (1993). *Prawda i racjonalność naukowa* [Truth and Scientific Rationality]. Kraków: Inter Esse.

Grodziński, E. (1969). *Język, metajęzyk, rzeczywistość* [Language, Metalanguage, Reality]. Warszawa: PWN.

Grzegorczyk, A. (1959). O pewnych formalnych konsekwencjach realizmu [On Some Formal Consequences of Realism]. In: *Fragmenty filozoficzne. Seria druga* [Philosophical Fragments. Second Series], pp. 7—14. Warszawa: PWN.

Grzegorczyk, A. (1993). *Życie jako wyzwanie* [Life as a Challenge]. Warszawa: Wydawnictwo IFiS PAN.

Hobbes, Th. ([1651] 1909). *Leviathan*. Oxford: Clarendon Press.

Hulewicz, J. (1924). W odpowiedzi Karolowi Irzykowskiemu. Zrozumialstwo [In Response to Karol Irzykowski. Intelligibility]. *Wiadomości Literackie* **1** (40—42), 2.

Ingarden, R. ([1925] 1972). O pytaniach esencjalnych [On Essential Questions]. In: *Z teorii języka i filozoficznych podstaw logiki* [From the Theory of Language and Philosophical Foundations of Logic), pp. 327—482. Warszawa: PWN.

Ingarden, R. ([1931] 1973). *The Literary Work of Art*. Evanston: Northwestern University Press.

Ingarden, R. (1934). Jeszcze «attuli mirohłady» [«Attuli mirohłady» Again]. *Wiadomości Literackie* **11** (43), 4.

Ingarden, R. ([1936] 1972). Analiza zdania warunkowego [Analysis of Conditional]. In: *Z teorii języka i filozoficznych podstaw logiki* [From the Theory of Language and Philosophical Foundations of Logic], pp. 260—270. Warszawa: PWN.

Ingarden, R. ([1947] 1964). *Der Streit um die Existenz der Welt*, vol. 1, Tübingen: Max Niemeyer Verlag.

Ingarden, R. ([1948] 1965). *Der Streit um die Existenz der Welt*, vol. 2(1—2), Tübingen, Max Niemeyer Verlag.

Ingarden, R. ([1949] 1957—1958). The Hypothetical Proposition. *Philosophy and Phenomenological Research* **18** (4), 435—450.

Ingarden, R. ([1964] 1969). Was wir über die Werte nicht wißen. In: *Erlebnis, Kunstwerk und Wert*, pp. 97—141. Max Niemeyer Verlag.

Irzykowski, K. (1924). Niezrozumialstwo [Unintelligibility]. *Wiadomości Literackie* **1** (38), 1.

Jadacka, H. (1976). *Termin techniczny — pojęcie, budowa, poprawność* [Technical Term — Notion, Structure, Correctness]. Warszawa: Wydawnictwo CT NOT.

Jadacki, J.J. (1980). On the Sources of Contemporary Polish Logic. *Dialectics and Humanism* **7** (4), 163—183.

Jadacki, J.J. (1982). On So-Called Theories of Truth. *Dialectics and Humanism* **9** (2), 145—149.

Jadacki, J.J. ([1985] 1996). O odpowiedzialności za słowo [On the Responsibility for Words]. In: *Metafizyka i semiotyka* [Metaphysics and Semiotics], pp. 244—258. Warszawa: Wydawnictwo WFiS UW.

Jung, C.G. ([1961] 1967). *Erinnerungen, Traüme, Gedanken*. Zürich: Rascher Verlag.

Kierkegaard, S. ([1850] 1933). *Einübung im Christentum*. Jena : Eugen Diederichs Verlag.

Kmita, J. (1964). Problem wartości logicznej ocen [The Problem of Truth-Value of Value-Judgements]. *Studia Filozoficzne* **7** (1), 119—139.

Kmita, J. (1966). Stwierdzanie a wyrażanie [Stating and Expressing]. *Studia Metodologiczne* **2**, 35—55.

Koj, L. (1971). Analiza pytań. I. Problem terminów pierwotnych logiki pytań [Analysis of Questions. I. The Problem of Primitive Terms of the Logic of Questions]. *Studia Semiotyczne* **2**, 99—113.

Kopania, J. (1987). Logika pytań [The Logic of Questions]. In: W. Marciszewski (ed.), *Logika formalna. Zarys encyklopedyczny* [Formal Logic. Encyclopaedic Outline], pp. 296—310. Warszawa: PWN.

Kotarbińska, J. (1964). Controversy on the Applicability Limits of Logical Methods. *Logique et Analyse* **8** (29), 1—29.

Kotarbiński, T. (1920). Sprawa istnienia przedmiotów idealnych [The Problem of Existence of Ideal Objects]. In: [Kotarbiński 1958], pp. 7—39.

Kotarbiński, T. ([1929] 1966). *Gnosiology. The Scientific Approach to the Theory of Knowledge.* Translated by O.A. Wojtasiewicz. Oxford & Wrocław: Pergamon Press & Ossolineum.

Kotarbiński, T. (1930—1931). Uwagi na temat reizmu [Notes on the Subject of Reism]. In: [Kotarbiński 1958], pp. 104—113.

Kotarbiński, T. (1931). O różnych znaczeniach słowa "materializm" [On Various Meanings of the Word "Materialism"]. In: [Kotarbiński 1958], pp. 69—89.

Kotarbiński, T. ([1934] 1960). The Concept of Action. *The Journal of Philosophy* **57** (7), 215—228.

Kotarbiński, T. ([1935] 1955). The Fundamental Ideas of Pansomatism". *Mind* **64**, 488—500.

Kotarbiński, T. (1949). The Reistic, or Concretistic, Approach. In: [Kotarbiński 1929], pp. 420—428.

Kotarbiński, T. (1954). On the Classification of Names. In: [Kotarbiński 1929], pp. 389—402.

Kotarbiński, T. (1956). The Concept of "Essence of Thing". In: [Kotarbiński 1929], pp. 410—419.

Kotarbiński, T. (1958). *Wybór pism* [Selected Writings], vol. 2. Warszawa: PWN.

Kraszewski, Z. (1972). *Metodologia filozofii* [Methodology of Philosophy]. Warszawa: PWN.

Krąpiec, M.A., S. Kamiński, Z.J. Zdybicka and P. Jaroszyński (1992). *Wprowadzenie do filozofii* [Introduction to Philosophy]. Lublin: Wydawnictwa KUL.

Kubiński, T. (1971). *Wstęp do logicznej teorii pytań* [Introduction to the Logical Theory of Questions]. Warszawa: PWN.

Lombroso, C. (1864). [Quotation from *The Man of Genius*.] In: [Cohen 1960], p. 236.

Łukasiewicz, J. (1906). Analiza i konstrukcja pojęcia przyczyny [Analysis and Construction of the Notion of Cause]. In: [Łukasiewicz 1961], pp. 9—62.

Łukasiewicz, J. (1907). Logika a psychologia [Logic and Psychology]. In: [Łukasiewicz 1961], pp. 63—65.

Łukasiewicz, J. (1912). Creative Elements in Science. In: [Łukasiewicz 1970], pp. 1—15.

Łukasiewicz, J. (1913). Zagadnienia prawdy [Problems of Truth]. In: *Logika i filozofia* [Logic and Philosophy], pp. 55—56. Warszawa: Wydawnictwo WFiS UW.

Łukasiewicz, J. (1934). On the History of the Logic of Propositions. In: [Łukasiewicz 1970], pp. 197—217.

Łukasiewicz, J. (1961). *Z zagadnień logiki i filozofii* [From the Problems of Logic and Philosophy]. Warszawa: PWN.

Łukasiewicz, J. (1970). *Selected Works.* Amsterdam & Warszawa: North-Holland Publishing Company & PWN.

Mallarmé, S. ([1887] 1982). "Cantique de saint Jean". In: *Les plus belles pages de la poésie français*, pp. 495—496. Paris: Sélection du Reader's Digest.

Marciszewski, W. (ed.) (1970a). *Mała encyklopedia logiki* [Little Encyclopaedia of Logic]. Wrocław: Ossolineum.

Marciszewski, W. (1970b). Funktor [Functor]. In: [Marciszewski 1970a], p. 73.

Marciszewski, W. (1970c). Kategoria semantyczna [Semantic Category]. In: [Marciszewski 1970a], pp. 99—101.

Marciszewski, W. (1970d). Nonsens [Nonsense]. In: [Marciszewski 1970a], pp. 190—191.

Marciszewski, W. (1970e). Prawda [Truth]. In: [Marciszewski 1970a], pp. 215—217.

Marciszewski, W. (1970f). Sensowność [Meaningfulness]. In: [Marciszewski 1970a], pp. 264.

Marciszewski, W. (1981). Categorial Grammar. In: W. Marciszewski (ed.), *Dictionary of Logic as Applied in the Study of Language. Concepts/Methods/Theories*, pp. 43—51. The Hague: Martinus Nijhoff Publishers,

Metallmann, J. (1939). *Wprowadzenie do zagadnień filozoficznych* [Introduction to Philosophical Problems], vol. 1. Kraków: D.E. Friedlein.

Moore, G.E. ([1953] 1958). *Some Main Problems of Philosophy*, London: George Allen and Unwin Ltd.

Nowak, L. (1970). O pojęciu wyrażania [On the Notion of Expressing]. *Studia Semiotyczne* **2**, 89— 97.

Nowak, L. (1977). *Wstęp do idealizacyjnej teorii nauki* [Introduction to Idealizational Theory of Science]. Warszawa: PWN.

Ossowska, M. ([1928] 1930). La fonction du langage d'exprimer nos idée. In: *Second Congrès Polonais de Philosophie. Varsovie 1927. Rapports et Comptes-Rendus*, pp. 134—136. Varsovie: *s.e.n.*

Ossowska, M. ([1931] 1983). Słowa i myśli [Words and Thoughts]. In: *O człowieku, moralności i nauce* [On Man, Morality and Science], pp. 183—225. Warszawa: PWN.

Pelc, J. (1982). Własności semiotyczne znaku [Semiotic Properties of Sign]. In: *Wstęp do semiotyki* [Introduction to Semiotics), pp. 227—233. Warszawa: Wiedza Powszechna. [Fragments as] Semiotic and Non-Semiotic Meaning, *American Journal of Semiotics* **1** (4), 1—19.

Pelc, J. (1985). Conditionals and Intensionality. In: H. Parret and H.-G. Ruprecht (eds.), *Aims and Prospects of Semiotics*, pp. 53—72. Amsterdam: John Benjamins B.V.

Poletyło, M. (1970). Pytanie [Question]. In: [Marciszewski 1970a], pp. 232—234.

Popper, K. ([1934] 1961). *The Logic of Scientific Discovery*. New York: Science Editions, Inc.

Presley, Ch.F. (1961). Arguments about Meaninglessness. *The British Journal of the Philosophy of Science* **12** (47), 225—234.

Przełęcki, M. (1969). *The Logic of Empirical Theories*. London: Routledge and Kegan Paul.

Przełęcki, M. (1981). On What There Is not. *Dialectics and Humanism* **8** (4), 123—129.

Przełęcki, M. (1996). *Poza granicami nauki* [Beyond the Limits of Science]. Warszawa: PTS.

Przełęcki, M. (1998). The Postulate of Precision: Its Sense and Its Limits. In: K. Kijania-Placek and J. Woleński (eds.), *The Lvov-Warsaw School and Contemporary Philosophy*, pp. 209—218. Dordrecht: Kluwer Academic Publishers.

Puzynina, J. (1992). *Język i wartości* [Language and Values]. Warszawa: PWN.

Quine, W.v.O. (1937). Logic and the Reification of Universals. In: [Quine 1953a], pp. 102—129.

Quine, W.v.O. ([1940] 1955). *Mathematical Logic*. Cambidge (Mass.): Harvard University Press.

Quine, W.v.O. (1950). Identity, Ostention, and Hypostasis. In: [Quine 1953a], pp. 65—79.

Quine, W.v.O. (1953a). *From a Logical Point of View*. Cambridge (Mass.): Harvard University Press.

Quine, W.v.O. (1953b). Meaning and Existential Inference. In: [Quine 1953a], pp. 160—167.

Quine, W.v.O. ([1966—1974] 1977). On Multiplying Entities. In: *"The Ways of Paradox" and Other Essays*, pp. 259—264. Cambridge (Mass.): Harvard University Press.

Reichenbach, H. ([1947] 1956). *Elements of Symbolic Logic*, New York: The Macmillan Company.

Rubczyński, W. (1906). Kryterium prawdy w teorii poznania pierwszych stoików i u Kartezjusza [Criterion of Truth in the Theory of Knowledge of the First Stoics and Descartes]. *Przegląd Filozoficzny* **9** (4), 358—368.

Russell, B. ([1931] 1954). *The Scientific Outlook*. London: George Allen and Unwin Ltd.

Russell, B. (1959). *My Philosophical Development*. London: George Allen and Unwin Ltd.

Salamucha, J. (1946). Z historii jednego wyrazu ("istota") [From the History of One Word ("Essence")]. In: *Wiedza i wiara* [Knowledge and Faith], pp. 71—82. Lublin: TN KUL.

Sarbiewski, M.K. ([1623] 1954). *De perfecta poesi*. Wrocław: Ossolineum.

Staff, L. ([1958] 1967). Gałczyński. In: *Poezje zebrane* [Collected Poetries], vol. 2, p. 946. Warszawa: PIW.

Stanosz, B. (1985). *Wprowadzenie do logiki formalnej. Podręcznik dla humanistów* [Introduction to Formal Logic. A Manual for Students of Humanities]. Warszawa: PWN.

Stanosz, B. and A. Nowaczyk (1976). *Logiczne podstawy języka* [Logical Foundations of Language]. Ossolineum: Wrocław.

Stonert, H. (1964a). Języki i teorie adekwatne z ontologią języka nauki [Languages and Theories Adequate to the Ontology of Scientific Language]. *Studia Logica* **15**, 49—77.

Stonert, H. (1964b). *Język i nauka* [Language and Science]. Warszawa: WP.

Stępień, A.B. (1964). *Wprowadzenie do metafizyki* [Introduction to Metaphysics]. Kraków: Znak.

Strawiński, W. (1991). *Prostota, redukcja, jedność nauki* [Simplicity, Reduction, Unity of Science]. Warszawa: Wydawnictwo FRA.

Strawson, P.F. (1959). *Individuals. An Essay in Descriptive Metaphysics*, London: Methuen and Co. Ltd.

Szaniawski, K. (1973). O pojęciu uznawania zdań [On the Notion of Acceptance of Sentences]. *Studia Filozoficzne* **16** (6), 189—197.

Szumilewicz, I. (1966). Prostota a prawda [Simplicity and Truth]. *Gdańskie Zeszyty Humanistyczne. Filozofia* **9** (2), 75—98.

Szymura, J. (1974). *Język, mowa i prawda w perspektywie fenomenologii lingwistycznej J.L. Austina* [Language, Speech, and Truth in the Perspective of J.L. Austin's Linguistic Phenomenology]. Wrocław: Ossolineum.

Talleyrand, Ch.-M. de (ca. 1800). [Quotation.] In: [Cohen 1960], p. 382.

Tarski, A. ([1933] 1956). The Concept of Truth in Formalized Languages. In: *Logic, Semantics, Metamathematics*, pp. 152—278. Oxford: Clarendon Press.

Tarski, A. ([1969] 1986). Truth and Proof. In: *Collected Papers*, vol. 4, pp. 401—423. Basel: Birkhäuser.

Tatarkiewiczowie, T. and W. (1979). *Wspomnienia* [Reminiscences]. Warszawa: PIW.

Tokarz, M. (1984). *Wprowadzenie do logiki* [Introduction to Logic]. Katowice: Uniwersytet Śląski.

Tolstoy, L. (1865—1869). [Quotation from *War and Peace*]. In: [Cohen 1960], p. 399.

Tomaszewski, T. (1936). *Geneza ocen niedorzeczności* [Genesis of Estimating Absurdities]. Lwów: Instytut Psychologii Uniwersytetu Jana Kazimierza.

Tuwim, J. ([1923] 1991). Słopiewnie. In: *Poezje* [Poetries], p. 144—146. Lublin: Wydawnictwo Lubelskie.

Twardowski, K. ([1894] 1977). *On the Content and Object of Presentations*. The Hague: Nijhoff.

Twardowski, K. (1901). *Zasadnicze pojęcia dydaktyki i logiki* [Fundamental Notions of Didactic and Logic]. Lwów: Towarzystwo Pedagogiczne.

Twardowski, K. (1919—1920). On Clear and Unclear Philosophical Style. In: [Twardowski 1999], p. 257—259.

Twardowski, K. (1925). Theory of Knowledge. In: [Twardowski 1999], p. 181—239.

Twardowski, K. (1999). *On Actions, Products and Other Topics in Philosophy. Poznań Studies in the Philosophy of the Sciences and the Humanities*, vol. 67. Amsterdam: Rodopi.

Ujejski, J. (1924). Postscriptum wydawcy [Postscript of Editor]. *Wiadomości Literackie* **1** (46), 2.

Williams, R. ([1958] 1971). [Quotation from *Culture and Society*.] In: J.M. and M.J. Cohen, *The Penguin Dictionary of Modern Quotations*, p. 242. Harmondsworth: Penguin Books Ltd.

Winner, Th.G. (1984). Mythic and Modern Elements in the Art of Ladislav Fuks: Natalia Mooshaber's Mice. In: J. Pelc *et. al.* (eds.), *Signs, System and Function*, pp. 457—477. Berlin: Mouton Publishers.

Witkiewicz, S.I. ([1927] 1976). Wstęp do rozważań nad «niezrozumialstwem» [Introduction to the Considerations Concerning «Unintelligibility»]. In: *"O znaczeniu filozofii dla krytyki" i inne artykuły polemiczne* ["On the Importance of Philosophy for Criticism" and Other Polemic Papers], pp. 361—366. Warszawa: PWN.

Wittgenstein, L. ([1922] 1933). *Tractatus Logico-Philosophicus*. London: Kegan Paul, Trech, Trubner and Co. Ltd.

Wolniewicz, B. (1968). *Rzeczy i fakty. Wstęp do pierwszej filozofii Wittgensteina* [Things and Facts. Introduction to the First Wittgenstein's Philosophy]. Warszawa: PWN.

Wolniewicz, B. (1978). Sytuacje jako korelaty semantyczne zdań [Situations as Semantic Correlates of Sentences]. *Studia Filozoficzne* **21** (2), 27—41.

Wolniewicz, B. (1986). Myśl Elzenberga [Elzenberg's Thought]. In: *Filozofia i wartości* [Philosophy and Values], pp. 69—94. Warszawa: WFiS UW.

Wolter, W. and M. Lipczyńska ([1973] 1980). *Elementy logiki* [Elements of Logic]. Warszawa: PWN.

Wójcicki, R. (1972). Metody formalne w problematyce teoriopoznawczej [Formal Methods in Epistemological Problems]. *Studia Filozoficzne* **15** (1), 13—42.

Wójcicki, T. (1982). *Wykłady z metodologii nauk* [Lectures on the Methodology of Sciences]. Warszawa: PWN.

INDEX OF NAMES

INDEX OF SUBJECTS

NB. The symbol "●" separates syntactic types of contexts that contain the main term of a given item.

E

POZNAŃ STUDIES IN THE PHILOSOPHY OF THE SCIENCES AND THE HUMANITIES

Contents of back issues

VOLUME 1 (1975)

Main topics:
The Method of Humanistic Interpretation; The Method of Idealization; The Reconstruction of Some Marxist Theories.
(sold out)

VOLUME 2 (1976)

Main topics:
Idealizational Concept of Science; Categorial Interpretation of Dialectics.
(sold out)

VOLUME 3 (1977)

Main topic:
Aspects of the Production of Scientific Knowledge.
(sold out)

VOLUME 4 (1978)

Main topic:
Aspects of the Growth of Science.
(sold out)

VOLUME 5 (1979)

Main topic:
Methodological Problems of Historical Research.
(sold out)

VOLUME 6 (1982)

SOCIAL CLASSES ACTION & HISTORICAL MATERIALISM

Main topics:
On Classes; On Action; The Adaptive Interpretation of Historical Materialism; Contributions to Historical Materialism.
(sold out)

VOLUME 7 (1982)

DIALECTICAL LOGICS FOR THE POLITICAL SCIENCE
(Edited by Hayward R. Alker, Jr.)

VOLUME 8 (1985)

CONSCIOUSNESS: METHODOLOGICAL AND PSYCHOLOGICAL APPROACHES
(Edited by Jerzy Brzeziński)

VOLUME 9 (1986)

THEORIES OF IDEOLOGY AND IDEOLOGY OF THEORIES
(Edited by Piotr Buczkowski and Andrzej Klawiter)

VOLUME 10 (1987)

WHAT IS CLOSER-TO-THE-TRUTH?
A PARADE OF APPROACHES TO TRUTHLIKENESS
(Edited by Theo A.F. Kuipers)

VOLUME 11 (1988)

NORMATIVE STRUCTURES OF THE SOCIAL WORLD
(Edited by Giuliano di Bernardo)

VOLUME 12 (1987)

POLISH CONTRIBUTIONS TO THE THEORY AND PHILOSOPHY OF LAW
(Edited by Zygmunt Ziembiński)

VOLUME 13 (1988)

DIMENSIONS OF THE HISTORICAL PROCESS
(Edited by Leszek Nowak)

VOLUME 14 (1990)

ACTION AND PERFORMANCE: MODELS AND TESTS.
CONTRIBUTIONS TO THE QUANTITATIVE PSYCHOLOGY
AND ITS METHODOLOGY
(Edited by Jerzy Brzeziński and Tadeusz Marek)

VOLUME 15 (1989)

VISIONS OF CULTURE AND THE MODELS OF CULTURAL SCIENCES
(Edited by Jerzy Kmita and Krystyna Zamiara)

VOLUME 16 (1990)

IDEALIZATION I: GENERAL PROBLEMS
(Edited by Jerzy Brzeziński, Francesco Coniglione, Theo A.F. Kuipers and
Leszek Nowak)

VOLUME 17 (1990)

IDEALIZATION II: FORMS AND APPLICATIONS
(Edited by Jerzy Brzeziński, Francesco Coniglione, Theo A.F. Kuipers and
Leszek Nowak)

VOLUME 18 (1990)

STUDIES ON MARIO BUNGE'S *TREATISE*
(Edited by Paul Weingartner and Georg J.W. Dorn)

VOLUME 19 (1990)

NARRATION AND EXPLANATION.
CONTRIBUTION TO THE METHODOLOGY OF THE HISTORICAL RESEARCH
(Edited by Jerzy Topolski)

VOLUME 20 (1990)

Jürgen Ritsert
MODELS AND CONCEPTS OF IDEOLOGY

VOLUME 21 (1991)

PROBABILITY AND RATIONALITY
STUDIES ON L. JONATHAN COHEN'S PHILOSOPHY OF SCIENCE
(Edited by Ellery Eells and Tomasz Maruszewski)

VOLUME 22 (1991)

THE SOCIAL HORIZON OF KNOWLEDGE
(Edited by Piotr Buczkowski)

VOLUME 23 (1991)

ETHICAL DIMENSIONS OF LEGAL THEORY
(Edited by Wojciech Sadurski)

VOLUME 24 (1991)

ADVANCES IN SCIENTIFIC PHILOSOPHY
ESSAYS IN HONOUR OF PAUL WEINGARTNER ON THE OCCASION OF THE 60TH ANNIVERSARY OF HIS BIRTHDAY
(Edited by Gerhard Schurz and Georg J.W. Dorn)

VOLUME 25 (1992)

IDEALIZATION III: APPROXIMATION AND TRUTH
(Edited by Jerzy Brzeziński and Leszek Nowak)

VOLUME 26 (1992)

IDEALIZATION IV: INTELLIGIBILITY IN SCIENCE
(Edited by Craig Dilworth)

VOLUME 27 (1992)

Ryszard Stachowski

THE MATHEMATICAL SOUL.
AN ANTIQUE PROTOTYPE OF THE MODERN MATEMATISATION OF PSYCHOLOGY

VOLUME 28 (1993)

POLISH SCIENTIFIC PHILOSOPHY:
THE LVOV-WARSAW SCHOOL
(Edited by Francesco Coniglione, Roberto Poli and Jan Woleński)

VOLUME 29 (1993)

Zdzisław Augustynek and Jacek J. Jadacki

POSSIBLE ONTOLOGIES

VOLUME 30 (1993)

GOVERNMENT: SERVANT OR MASTER?
(Edited by Gerard Radnitzky and Hardy Bouillon)

VOLUME 31 (1993)

CREATIVITY AND CONSCIOUSNESS.
PHILOSOPHICAL AND PSYCHOLOGICAL DIMENSIONS
(Edited by Jerzy Brzeziński, Santo Di Nuovo, Tadeusz Marek and Tomasz Maruszewski)

VOLUME 32 (1993)

FROM ONE-PARTY-SYSTEM TO DEMOCRACY
(Edited by Janina Frentzel-Zagórska)

J. Frentzel-Zagórska, *Introduction*. **Part I: Theoretical Approaches** – Z. Bauman, *A Post-modern Revolution*; L. Holmes, *On Communism, Post-communism, Modernity and Post-modernity*; L. Nowak, *The Totalitarian Approach and the History of Socialism*; J. Pakulski, *East European Revolutions and 'Legitimacy Crisis'*. **Part II: The Transitional Period** – A. Czarnota, M. Krygier, *From State to Legal Traditions? Prospects for the Rule of Law after Communism*; M. Szabó, *Social Protest in a Post-communist Democracy: Taxi Drivers' Demonstration in Hungary*; Z. Bauman, *Dismantling Patronage State*; E. Mokrzycki, *Between Reform and Revolution: Eastern Europe Two Years after the Fall of Communism*; J. Frentzel-Zagórska, *The Road to Democratic Political System in Post-communist Eastern Europe*. **Part III: The Case of Yugoslavia** – R.F. Miller, *Yugoslavia: The End of the Experiment*.

VOLUME 33 (1993)

SOCIAL SYSTEM, RATIONALITY, AND REVOLUTION
(Edited by Marcin Paprzycki and Leszek Nowak)

Introduction. **On the Nature of Social System** – U. Preuss, *Political Order and Democracy. Carl Schmitt and his Influence*; K. Paprzycka, *A Paradox in Hobbes' Philosophy of Law*; S.

Esquith, *Democratic Political Dialogue*; E. Jeliński, *Democracy in Polish Reformist Socialist Thought*; K. Paprzycka, *The Master and Slave Configuration in Hegel's System*; M. Godelier, *Lévi-Strauss, Marx and After. A Reappraisal of Structuralist and Marxist Tools for Social Logics*; K. Niedźwiadek, *On the Structure of Social System*; W. Czajkowski, *Social Being and Its Reproduction*. **On Rationality and Captivity** – M. Ziółkowski, *Power and Knowledge*, L. Nowak, *Two Inter-Human Limits to the Rationality of Man*; M. Paprzycki, *The Non-Christian Model of Man: An Attempt at Psychological Explanation*; R. Egiert, *Toward the Sophisticated Rationalistic Model of Man*. **On Social Revolution** – L. Nowak, *Revolution is an Opaque Progress but a Progress Nonetheless*; K. Paprzycka, M. Paprzycki, *How do Enslaved People Make Revolutions?*; G. Tomczak, *Is it Worth Winning a Revolution?*; K. Brzechczyn, *Civil Loops and the Absorption of Elites*; R. McCleary, *What Makes Marxist Historical Materialism Objective?*; G. Kotlarski, *Classes and Masses in Social Philosophy of Rosa Luxemburg*. **On Real Socialism** – E. Gellner, *The Civil and the Sacred*; W. Marciszewski, *Economics and the Idea of Information. Why Socialism must have Collapsed?*; L. Nowak, K. Paprzycka, M. Paprzycki, *On Multilinearity of Socialism*; A. Siegel, *The Overrepression Cycle in the Soviet Union. An Operationalization of a Theoretical Model*; K. Brzechczyn, *The State of the Teutonic Order as a Socialist Society*. **Discussions** – R. McCleary, *Socioanalysis and Philosophy*; W. Heller, *Methodological Remarks on the Public and the Private in Hannah Arendt's Political Philosophy*; K. Brzechczyn, *On Unsuccessful Conquest and Successful Subordination*.

VOLUME 34 (1994)

Izabella Nowakowa

IDEALIZATION V: THE DYNAMICS OF IDEALIZATIONS

Introduction; Chapter I: *Idealization and Theories of Correspondence*; Chapter II: *Dialectical Correspondence of Scientific Laws*; Chapter III: *Dialectical Correspondence in Science: Some Examples*; Chapter IV: *Dialectical Correspondence of Scientific Theories*; Chapter V: *Generalizations of the Rule of Correspondence*; Chapter VI: *Extensions of the Rule of Correspondence*; Chapter VII: *Correspondence and the Empirical Environment of a Theory*; Chapter VIII: *Some Methodological Problems of Dialectical Correspondence*.

VOLUME 35 (1993)

EMPIRICAL LOGIC AND PUBLIC DEBATE.
ESSAYS IN HONOUR OF ELSE M. BARTH
(Edited by E.C.W. Krabbe, R.J. Dalitz, and P.A. Smit)

Part I: Interpersonal Reasoning: Conflicts and Fallacies – T. Govier, *Needing Each Other for Knowledge: Reflections on Trust and Testimony*; J. Woods, *'Secundum quid' as a Research Programme*; G. Nuchelmans, *On the Fourfold Root of the 'argumentum ad hominem'*; F.H. van Eemeren, R. Grootendorst, *The History of the 'argumentum ad hominem' Since the Seventeenth Century*; L.S. van Epenhuysen, *Debate in a Bermuda Triangle of Medical Ethics*; E.C.W. Krabbe, *Reasonable Argument and Fallacies in the Kok-Stekelenburg Debate*; **Part II: Linguistic and Conceptual Tools** – J.D. North, *Some Weak Links in the Great Chain of Being*; A. Næss, *'You assert this?' An Empirical Study of Weight Expressions*; R. Wiche, *Gerrit Mannoury on the Communicative Functions of Negation in Ordinary Language*; J. Hoepelman, T. van Hoof, *Default and Dogma*; Ch. Goossens, *On the Logic of Nonmoral Commitment*; W. Marciszewski, *Arguments Founded on Creative Definitions*; R. Jorna, *Cognitive Science and Connectivism: Friend and Enemy or Move and Counter-Move, an Application of Empirical Logic*; **Part III: Dialectical Climates and Tempests** – R.H.

Johnson, *Dialectical Fields and Manifest Rationality*; P. du Preez, *Reason Which Cannot Be Reasoned With: What Is Public Debate and How Does it Change?*; M.A. Finocchiaro, *Logic, Democracy, and Mosca*; P.A. Smit, *The Logic of Virtue and Terror*; G. van Benthem van den Bergh, *On Obstacles to Public Debate*; J.P. van Bendegem, *Real-Life Mathematics versus Ideal Mathematics: The Ugly Truth*; **Part IV: The Disempowernment of Woman: Strategies and Counter-Moves** – V. Songe-Møller, *The Road of Being and the Exclusion of the Feminine. An Analysis of the Poem of Parmenides*; R.J. Dalitz, *The Subjection of Women in the Contractual Society. An Analysis of Thomas Hobbes' Theory of Agreement*; H. Schröder, *Anti-Semitism and anti-Feminism Again: The Dissemination of Otto Weininger's* Sex and Character *in the Seventies and Eighties*; J.R. Richards, *Traditional Spheres and Traditional Logic*.

VOLUME 36 (1994)

MARXISM AND COMMUNISM: POSTHUMOUS REFLECTIONS ON POLITICS, SOCIETY, AND LAW
(Edited by Martin Krygier)

M. Krygier, *Introduction*; A. Flis, *From Marx to Real Socialism: The History of a Utopia*; P. Marciniak, *The Collapse of Communism: Defeat or Opportunity for Marxism in Eastern Europe*; J. Clark, A. Wildavsky, *Chronicle of a Collapse Foretold: How Marx Predicted the Demise of Communism (Although He Called It "Capitalism")*; L. Nowak, *Political Theory and Socialism. On the Main Paradigms of Political Power and Their Methodological and Historical Legitimation*; E. Mokrzycki, *Marxism, Sociology, and "Real Socialism"*; R. Bäcker, *The Collapse of Communism and Theoretical Models*; A. Zybertowicz, *Three Deaths an Ideology: The Withering Away of Marxism and the Collapse of Communism. The Case of Poland*; M. Krygier, *Marxism, Communism, and the Rule of Law*; A. Czarnota, *Marxism, Ideology, and Law*; G. Skąpska, *The Legacy of Anti-Legalism*; A. Sajo, *Law and the Legal Scholarship in the Happiest Barrack and Among the Hungry Liberated: Personal Recollections*.

VOLUME 37 (1994)

THE SOCIAL PHILOSOPHY OF AGNES HELLER
(Edited by John Burnheim)

J. Burnheim, *Introduction*; M. Vajda, *A Lover of Philosophy – A Lover of Europe*; P. Despoix, *On the Possibility of a Philosophy of Values. A Dialogue within the Budapest School*; M. Jay, *Women in Dark Times: Agnes Heller and Hannah Arendt*; J.P. Arnason, *The Human Condition and the Modern Predicament*; R.J. Bernstein, *Agnes Heller: Philosophy, Rational Utopia and Praxis*; Z. Bauman, *Narrating Modernity*; P. Beilharz, *Theories of History – Agnes Heller and R.G. Collingwood*; P. Wolin, *Heller's Theory of Everyday Life*; P. Harrison, *Radical Philosophy and the Theory of Modernity*; A.J. Jacobson, *The Limits of Formal Justice*; P. Murphy, *Civility and Radicalism*; P. Murphy, *Pluralism and Politics*; V. Camps, *The Good Life: A Moral Gesture*; L. Boella, *Philosophy Beyond the Baseless and Tragic Character of Action*; G. Márkus, *The Politics of Morals*; A. Heller, *A Reply to My Critics*; *The Bibliography of Agnes Heller*.

VOLUME 38 (1994)

IDEALIZATION VI: IDEALIZATION IN ECONOMICS
(Edited by Bert Hamminga and Neil B. De Marchi)

Introduction – B. Hamminga, N. De Marchi, *Preface*; B. Hamminga, N. De Marchi, *Idealization and the Defence of Economics: Notes Toward a History.* **Part I: General Observations on Idealization in Economics** – K.D. Hoover, *Six Queries about Idealization in an Empirical Context*; B. Walliser, *Three Generalization Processes for Economic Models*; S. Cook, D. Hendry, *The Theory of Reduction in Econometrics*; M.C.W. Janssen, *Economic Models and Their Applications*; A.G. de la Sienra, *Idealization and Empirical Adequacy in Economic Theory*; I. Nowakowa, L. Nowak, *On Correspondence between Economic Theories*; U. Mäki, *Isolation, Idealization and Truth in Economics.* **Part II: Case Studies of Idealization in Economics** – N. Cartwright, *Mill and Menger: Ideal Elements and Stable Tendencies*; W. Balzer, *Exchange Versus Influence: A Case of Idealization*; K. Cools, B. Hamminga, T.A.F. Kuipers, *Truth Approximation by Concretization in Capital Structure Theory*; D.M. Hausman, *Paul Samuelson as Dr. Frankenstein: When an Idealization Runs Amuck*; H.A. Keuzenkamp, *What if an Idealization is Problematic? The Case of the Homogeneity Condition in Consumer Demand*; W. Diederich, *Nowak on Explanation and Idealization in Marx's Capital*; G. Jorland, *Idealization and Transformation*; J. Birner, *Idealizations and Theory Development in Economics. Some History and Logic of the Logic Discovery.* **Discussions** – L. Nowak, *The Idealizational Methodology and Economics. Replies to Diederich, Hoover, Janssen, Jorland and Mäki.*

VOLUME 39 (1994)

PROBABILITY IN THEORY-BUILDING.
EXPERIMENTAL AND NON-EXPERIMENTAL APPROACHES
TO SCIENTIFIC RESEARCH IN PSYCHOLOGY
(Edited by Jerzy Brzeziński)

Part I: Probability and the Idealizational Theory of Science – M. Gaul, *Statistical Dependencies, Statements and the Idealizational Theory of Science.* **Part II: Probability – Theoretical Concepts in Psychology – Measurement** – D. Wahlstein, *Probability and the Understanding of Individual Differences*; B. Krause, *Modeling Cognitive Learning Steps*; D. Heyer, R. Mausfeld, *A Theoretical and Experimental Inquiry into the Relation of Theoretical Concepts and Probabilistic Measurement Scales in Experimental Psychology.* **Part III: Methods of Data Analysis** – T.B. Iwiński, *Rough Set Methods in Psychology*; W. Koutstaal, R. Rosenthal, *Contrast Analysis in Behavioral Research.* **Part IV: Artifacts in Psychological Research and Diagnostic Assessment** – D.B. Strohmetz, R.L. Rosnow, *A Mediational Model of Research Artifacts*; J. Brzeziński, *Dimensions of Diagnostic Space.*

VOLUME 40 (1995)

THE HERITAGE OF KAZIMIERZ AJDUKIEWICZ
(Edited by Vito Sinisi and Jan Woleński)

Preface; J. Giedymin, *Ajdukiewicz's Life and Personality*; K. Ajdukiewicz, *My Philosophical Ideas*; L. Albertazzi, *Some Elements of Transcendentalism in Ajdukiewicz*; T. Batóg, *Ajdukiewicz and the Development of Formal Logic*; A. Church, *A Theory of the Meaning of Names*; M. Czarniawska, *The Way from Concept to Thought. Does it Exist in Ajdukiewicz's Semantical Theory?*; E. Dölling, *Real Objects and Existence*; J. Giedymin, *Radical Conventionalism, Its Background and Evolution: Poincaré, Le Roy, Ajdukiewicz*; J.J. Jadacki,

Definition, Explication and Paraphrase in the Ajdukiewiczian Tradition; G. Küng, *Ajdukiewicz's Contribution to the Realism/Idealism Debate*; J. Maciaszek, *Problems of Proper Names: Ajdukiewicz and Some Contemporary Results*; W. Marciszewski, *Real Definitions and Creativity*; K. Misiuna, *Categorial Grammar and Ontological Commitment*; L. Nowak, *Ajdukiewicz and the Status of the Logical Theory of Natural Language*; A. Olech, *Several Remarks on Ajdukiewicz's and Husserl's Approaches to Meaning*; A. Orenstein, *Existence Sentences*; J. Pasek, *Ajdukiewicz on Indicative Conditionals*; R. Poli, *The Problem of Position: Ajdukiewicz and Leibniz on Intensional Expressions*; M. Przełęcki, *The Law of Excluded Middle and the Problem of Idealism*; H. Skolimowski, *Ajdukiewicz, Rationality and Language*; K. Szaniawski, *Ajdukiewicz on Non-deductive Inference*; A. Varzi, *Variable-Binders as Functors*; V. Vasyukov, *Categorial Semantics for Ajdukiewicz-Lambek Calculus*; H. Wassing, *On the Expressiveness of Categorial Grammar*; J. Woleński, *On Ajdukiewicz's Refutation of Scepticism*; *Bibliography of Ajdukiewicz; Bibliography on Ajdukiewicz (writings in western languages)*.

VOLUME 41 (1994)

HISTORIOGRAPHY BETWEEN MODERNISM AND POSTMODERNISM.
CONTRIBUTIONS TO THE METHODOLOGY OF
THE HISTORICAL RESEARCH
(Edited by Jerzy Topolski)

Editor's Introduction; J. Topolski, *A Non-postmodernist Analysis of Historical Narratives*; F.R. Ankersmit, *Historism, Postmodernism and Historiography*; D. Carr, *Getting the Story Straight: Narrative and Historical Knowledge*; W. Wrzosek, *The Problem of Cultural Imputation in History. Culture versus History*; J. Tacq, *Causality and Virtual Finality*; G. Zalejko, *Soviet Historiography as a "Normal Science"*; H. Mamzer and J. Ostoja-Zagórski, *Deconstruction of Evolutionist Paradigm in Archaeology*; N. Lautier, *At the Crossroad of Epistemology and Psychology: Prospects of a Didactic of History*; Teresa Kostyrko, *Remarks on "Aesthetization" in Science on the Basis of History*.

VOLUME 42 (1995)

IDEALIZATION VII: IDEALIZATION, STRUCTURALISM,
AND APPROXIMATION
(Edited by Martti Kuokkanen)

Idealization, Approximation and Counterfactuals in the Structuralist Framework – T.A.F. Kuipers, *The Refined Structure of Theories*; C.U. Moulines and R. Straub, *Approximation and Idealization from the Structuralist Point of View*; I.A. Kieseppä, *A Note on the Structuralist Account of Approximation*; C.U. Moulines and R. Straub, *A Reply to Kieseppä*; W. Balzer and G. Zoubek, *Structuralist Aspects of Idealization*; A. Ibarra and T. Mormann, *Counterfactual Deformation and Idealization in a Structuralist Framework*; I.A. Kieseppä, *Assessing the Structuralist Theory of Verisimilitude*. **Idealization, Approxima-tion and Theory Formation** – L. Nowak, *Remarks on the Nature of Galileo's Methodological Revolution*; I. Niiniluoto, *Approximation in Applied Science*; E. Heise, P. Gerjets and R. Westermann, *Idealized Action Phases. A Concise Rubicon Theory*; K.G. Troitzsch, *Modelling, Simulation, and Structuralism*; V. Rantala and T. Vadén, *Idealization in Cognitive Science. A Study in Counterfactual Correspondence*; M. Sintonen and M. Kiikeri, *Idealization in Evolutionary Biology*; T. Tuomivaara, *On Idealizations in Ecology*; M. Kuokkanen and M. Häyry, *Early Utilitarianism and Its Idealizations from a Systematic Point of View*. **Idealization, Approximation and Measurement** – R. Westermann, *Measurement-Theoretical Idealizations and Empirical Research Practice*; U. Konerding, *Probability as an Idealization of Relative*

Frequency. A Case Study by Means of the BTL-Model; R. Suck and J. Wienöbst, *The Empirical Claim of Probality Statements, Idealized Bernoulli Experiments and their Approximate Version*; P.J. Lahti, *Idealizations in Quantum Theory of Measurement.*

VOLUME 43 (1995)

Witold Marciszewski and Roman Murawski
MECHANIZATION OF REASONING IN A HISTORICAL PERSPECTIVE

Chapter 1: *From the Mechanization of Reasoning to a Study of Human Intelligence*; Chapter 2: *The Formalization of Arguments in the Middle Ages*; Chapter 3: *Leibniz's Idea of Mechanical Reasoning at the Historical Background*; Chapter 4: *Between Leibniz and Boole: Towards the Algebraization of Logic*; Chapter 5: *The English Algebra of Logic in the 19th Century*; Chapter 6: *The 20th Century Way to Formalization and Mechanization*; Chapter 7: *Mechanized Deduction Systems.*

VOLUME 44 (1995)

THEORIES AND MODELS IN SCIENTIFIC PROCESSES
(Edited by William Herfel, Władysław Krajewski,
Ilkka Niiniluoto and Ryszard Wójcicki)

Introduction; **Part 1. Models in Scientific Processes** – J. Agassi, *Why there is no Theory of Models?*; M. Czarnocka, *Models and Symbolic Nature of Knowledge*; A. Grobler, *The Representational and the Non-Representational in Models of Scientific Theories*; S. Hartmann, *Models as a Tool for the Theory Construction: Some Strategies of Preliminary Physics*; W. Herfel, *Nonlinear Dynamical Models as Concrete Construction*; E. Kałuszyńska, *Styles of Thinking*; S. Psillos, *The Cognitive Interplay Between Theories and Models: the Case of 19th Century Optics.* **Part 2. Tools of Science** – N.D. Cartwright, T. Shomar, M. Suarez, *The Tool-Box of Science*; J. Echeverria, *The Four Contexts of Scienctific Activity*; K. Havas, *Continuity and Change; Kinds of Negation in Scientific Progress*; M. Kaiser, *The Independence of Scientific Phenomena*; W. Krajewski, *Scientific Meta-Philosophy*; I. Niiniluoto, *The Emergence of Scientific Specialties: Six Models*; L. Nowak, *Antirealism, (Supra-) Realism and Idealization*; R.M. Nugayev, *Classic, Modern and Postmodern Scientific Unification*; V. Rantala, *Translation and Scientific Change*; G. Schurz, *Theories and Their Applications – a Case of Nonmonotonic Reasoning*; W. Strawiński, *The Unity of Science Today*; V. Torosian, *Are the Ethics and Logic of Science Compatible?* **Part 3. Unsharp Approaches in Science** – E.W. Adams, *Problems and Prospects in a Theory of Inexact First-Order Theories*; W. Balzer, G. Zoubek, *On the Comparision of Approximative Empirical Claims*; G. Cattaneo, M. Luisa Dalla Chiara, R. Giuntini, *The Unsharp Approaches to Quantum Theory*; T.A.F. Kuipers, *Falsification Versus Effcient Truth Approximation*; B. Lauth, *Limiting Decidability and Probability*; J. Pykacz, *Many-Valued Logics in Foundations of Quantum Mechanics*; R.R. Zapatrin, *Logico-Algebraic Approach to Spacetime Quantization.*

VOLUME 45 (1995)

COGNITIVE PATTERNS IN SCIENCE AND COMMON SENSE.
GRONINGEN STUDIES IN PHILOSOPHY OF SCIENCE,
LOGIC, AND EPISTEMOLOGY
(Edited by Theo A.F. Kuipers and Anne Ruth Mackor)

L. Nowak, *Foreword*; **General Introduction** – T.A.F. Kuipers and A.R. Mackor, *Cognitive Studies of Science and Common Sense*; **Part I: Conceptual Analysis in Service of Various Research Programmes** – H. Zandvoort, *Concepts of Interdisciplinarity and Environmental Science*; R. Vos, *The Logic and Epistemology of the Concept of Drug and Disease Profile*; R.C. Looijen, *On the Distinction Between Habitat and Niché, and Some Implications for Species' Differentiation*; G.J. Stavenga, *Cognition, Irreversibility and the Direction of Time*; R. Dalitz, *Knowledge, Gender and Social Bias*; **Part II: The Logic of the Evaluation of Arguments, Hypotheses, Rules, and Interesting Theorems** – E.G.W. Krabbe, *Can We Ever Pin One Down to a Formal Fallacy?*; T.A.F. Kuipers, *Explicating the Falsificationist and Instrumentalist Methodology by Decomposing the Hypothetico-Deductive Method*; A. Keupink, *Causal Modelling and Misspecification: Theory and Econometric Historical Practice*; M.C.W. Janssen and Y.-H. Tan, *Default Reasoning and Some Applications in Economics*; B. Hamminga, *Interesting Theorems in Economics*; **Part III: Three Challenges to the Truth Approximation Programme** – S.D. Zwart, *A Hidden Variable in the Discussion About 'Language Dependency' of Truthlikeness*; H. Hettema and T.A.F. Kuipers, *Sommerfeld's* Atombau: *A Case Study in Potential Truth Approximation*; R. Festa, *Verisimilitude, Disorder, and Optimum Prior Probabilities*; **Part IV: Explicating Psychological Intuitions** – A.R. Mackor, *Intentional Psychology is a Biological Discipline*; J. Peijnenburg, *Hempel's Rationality. On the Empty Nature of Being a Rational Agent*; L. Guichard, *The Causal Efficacy of Propositional Attitudes*; M. ter Hark, *Connectionism, Behaviourism and the Language of Thought.*

VOLUME 46 (1996)

POLITICAL DIALOGUE: THEORIES AND PRACTICE
(Edited by Stephen L. Esquith)

Introduction – S.L. Esquith, *Political Dialogue and Political Virtue.* **Part I: The Modern Clasics** – A.J. Damico, *Reason's Reach: Liberal Tolerance and Political Discourse*; T.R. Machan, *Individualism and Political Dialogue*; R. Kukla, *The Coupling of Human Souls: Rousseau and the Problem of Gender Relations*; D.F. Koch, *Dialogue: An Essay in the Instrumentalist Tradition.* **Part II: Toward a Democratic Synthesis** – E. Simpson, *Forms of Political Thinking and the Persistence of Practical Philosophy*; J.B. Sauer, *Discoursee, Consensus, and Value: Conversations about the Intelligible Relation Between the Private and Public Spheres*; M. Kingwell, *Phronesis and Political Dialogue*, R.T. Peterson, *Democracy and Intellectual Mediation – After Liberalism and Socialism.* **Part III: Dialogue in Practice** – S. Rohr Scaff, L.A. Scaff, *Political Dialogue in the New Germany: The Burdens of Culture and an Asymmetrical Past*; J.H. Read, *Participation, Power, and Democracy*; S.E. Bennett, B. Fisher, D. Resnick, *Speaking of Politics in the United States: Who Talks to Whom, Why, and Why Not*; J. Forester, *Beyond Dialogue to Transformative Learning: How Deliberative Rituals Encourage Political Judgement in Community Planning Processes*; A. Fatić, *Retribution in Democracy.*

VOLUME 47 (1996)

EPISTEMOLOGY AND HISTORY. HUMANITIES AS A PHILOSOPHICAL PROBLEM AND JERZY KMITA'S APPROACH TO IT
(Edited by Anna Zeidler-Janiszewska)

A. Zeidler-Janiszewska, *Preface.* **Humanistic Knowledge** – K.O. Apel, *The Hermeneutic Dimension of Social Science and its Normative Foundation*; M. Czerwiński, *Jerzy Kmita's Epistemology*; L. Witkowski, *The Frankfurt School and Structuralism in Jerzy Kmita's Analysis*; A. Szahaj, *Between Modernism and Postmodernism: Jerzy Kmita's Epistemology*; A. Grzegorczyk, *Non-Cartesian Coordinates in the Contemporary Humanities*; A. Pałubicka, *Pragmatist Holism as an Expression of Another Disenchantment of the World*; J. Sójka, *Who is Afraid of Scientism?*; P. Ozdowski, *The Broken Bonds with the World*; J. Such, *Types of Determination vs. the Development of Science in Historical Epistemology*; P. Zeidler, *Some Issues of Historical Epistemology in the Light of the Structuralist Philosophy of Science*; M. Buchowski, *Via Media: On the Consequences of Historical Epistemology for the Problem of Rationality*; B. Kotowa, *Humanistic Valuation and Some Social Functions of the Humanities.* **On Explanation and Humanistic Interpretation** – T.A.F. Kuipers, *Explanation by Intentional, Functional, and Causal Specification*; E. Świderski, *The Interpretational Paradigm in the Philosophy of the Human Sciences*; L. Nowak, *On the Limits of the Rationalistic Paradigm*; F. Coniglione, *Humanistic Interpretation between Hempel and Popper*; Z. Ziembiński, *Historical Interpretation vs. the Adaptive Interpretation of a Legal Text*; W. Mejbaum, *Explaining Social Phenomena*; M. Ziółkowski, *The Functional Theory of Culture and Sociology*; K. Zamiara, *Jerzy Kmita's Social-Regulational Theory of Culture and the Category of Subject*; J. Brzeziński, *Theory and Social Practice. One or Two Psychologies?*; Z. Kwieciński, *Decahedron of Education (Components and aspects). The Need for a Comprehensive Approach*; J. Paśniczek, *The Relational vs. Directional Concept of Intentionality.* **The Historical Dimension of Culture and its Studies** – J. Margolis, *The Declension of Progressivism*; J. Topolski, *Historians Look at Historical Truth*; T. Jerzak-Gierszewska, *Three Types of the Theories of Religion and Magic*; H. Paetzold, *Mythos und Moderne in der Philosophie der symbolishen Formen Ernst Cassirers*; M. Siemek, *Sozialphilosophische Aspekte der Übersetzbarkeit.* **Problems of Artistic Practice and Its Interpretation** – S. Morawski, *Theses on the 20th Century Crisis of Art and Culture*; A. Erjavec, *The Perception of Science in Modernist and Postmodernist Artistic Practice*, G. Dziamski, *The Avant-garde and Contemporary Artistic Consciousness*; H. Orłowski, *Generationszugehörigkeit und Selbsterfahrung von (deutschen) Schriftstellern*; T. Kostyrko, *The "Transhistoricity" of the Structure of Work of Art and the Process of Value Transmission in Culture*; G. Banaszak, *Musical Culture as a Configuration of Subcultures*; A. Zeidler-Janiszewska, *The Problem of the Applicability of Humanistic Interpretation in the Light of Contemporary Artistic Practice*; J. Kmita, *Towards Cultural Relativism "with a Small 'r'"*; The Bibliography of Jerzy Kmita.

VOLUME 48 (1996)

THE SOCIAL PHILOSOPHY OF ERNEST GELLNER
(Edited by John A. Hall and Ian Jarvie)

J.A. Hall, I. Jarvie, *Preface*; J.A. Hall, I. Jarvie, *The Life and Times of Ernest Gellner.* **Part 1: Intelectual Background** – J. Musil, *The Prague Roots of Ernest Gellner's Thinking*; Ch. Hahn, *Gellner and Malinowski: Words and Things in Central Europe*; T. Dragadze, *Ernest Gellner in Soviet East.* **Part 2: Nations and Nationalism** – B. O'Leary, *On the Nature of Nationalism: An Appraisal of Ernest Gellner's Writings on Nationalism*; K. Minogue, *Ernest Gellner and the Dangers of Theorising Nationalism*; A.D. Smith, *History and Modernity:*

Reflection on the Theory of Nationalism; M. Mann, *The Emergence of Modern European Nationalism*; N. Stagardt, *Gellner's Nationalism: The Spirit of Modernisation?* **Part 3: Patterns of Development** – P. Burke, *Reflections on the History of Encyclopaedias*; A. MacFarlane, *Ernest Gellner and the Escape to Modernity*; R. Dore, *Soverein Individuals*; S. Eisenstadt, *Japan: Non Axial Modernity*; M. Ferro, *L'Indépendance Telescopée: De la Décolonisation a L'Impérialisme Multinational.* **Part 4: Islam** – A. Hammoudi, *Segmentarity, Social stratification, Political Power and Sainthood: Reflections on Gellner's Theses*; H. Munson, Jr., *Rethinking Gellner's Segmentary Analysis of Morocco's Ait^cAtta*; J. Baechler, *Sur le charisme*; Ch. Lindholm, *Despotism and Democracy: State and Society in PreModern Middle East*; H. Munson, *Jr. Muslism and Jew in Morocco: Reflections on the Distinction between Belief and Behaviour*; T. Asad, *The Idea of an Anthropology of Islam.* **Part 5: Science and Disenchantment** – P. Anderson, *Science, Politics, Enchantment*; R. Schroeder, *From the Big Divide to the Rubber Cage: Gellner's Conception of Science and Technology*; J. Davis, *Irrationality in Social Life.* **Part 6: Relativism and Universals** – J. Skorupski, *The Post-Modern Hume: Ernest Gellner's 'Enlightenment Fundamentalism'*; J. Wettersten, *Ernest Gellner: A Wittgensteinian Rationalist*; I. Jarvie, *Gellner's Positivism*; R. Boudon, *Relativising Relativism: When Sociology Refutes the Sociology of Science*; R. Aya, *The Empiricist Exorcist.* **Part 7: Philosophy of History** – W. McNeill, *A Swang Song for British Liberalism?*; A. Park, *Gellner and the Long Trends of History*; E. Leone, *Marx, Gellner, Power*; R. Langlois, *Coercion, Cognition and Production: Gellner's Challenge to Historical Materialism and Post-Modernism*; E. Gellner, *Reply to my Critics*; I. Jarvie, *Complete Bibliography of Gellner's Work.*

VOLUME 49 (1996)

THE SIGNIFICANCE OF POPPER'S THOUGHT
(Edited by Stefan Amsterdamski)

Karl Popper's Three Worlds – J. Watkins, *World 1, World 2 and the Theory of Evolution*; A. Grobler, *World 3 and the Cunning of Reason.* **The Scientific Method as Ethics** – J. Agassi, *Towards Honest Public Relations of Science*; S. Amsterdamski, *Between Relativism and Absolutism: the Popperian Ideal of Knowledge.* **The Open Society and its Prospects** – E. Gellner, *Karl Popper – The Thinker and the Man*; J. Woleński, *Popper on Prophecies and Predictions.*

VOLUME 50 (1996)

THE IDEA OF THE UNIVERSITY
(Edited by Jerzy Brzeziński and Leszek Nowak)

Introduction – K. Twardowski, *The Majesty of the University.* **I** – Z. Ziembiński, *What Can be Saved of the Idea of the University?*; L. Kołakowski, *What Are Universities for?*; L. Gumański, *The Ideal University and Reality*; Z. Bauman, *The Present Crisis of the Universities.* **II** – K. Ajdukiewicz, *On Freedom of Science*; H. Samsonowicz, *Universities and Democracy*; J. Topolski, *The Commonwealth of Scholars and New Conceptions of Truth*; K. Szaniawski, *Plus ratio quam vis.* **III** – L. Koj, *Science, Teaching and Values*; K. Szaniawski, *The Ethics of Scientific Criticism*; J. Brzeziński, *Ethical Problems of Research Work of Psychologists.* **IV** – J. Goćkowski, *Tradition in Science*; J. Kmita, *Is a "Creative Man of Knowledge" Needed in University Teaching?*; L. Nowak, *The Personality of Researchers and the Necessity of Schools in Science.* **Recapitulation** – J. Brzeziński, *Reflections on the University.*

VOLUME 51 (1997)

KNOWLEDGE AND INQUIRY: ESSAYS ON JAAKKO HINTIKKA'S EPISTEMOLOGY AND PHILOSOPHY OF SCIENCE
(Edited by Matti Sintonen)

M. Sintonen, *From the Science of Logic to the Logic of Science.* **I: Historical Perspectives** – Z. Bechler, *Hintikka on Plentitude in Aristotle*; M.-L. Kakkuri-Knuuttila, *What Can the Sciences of Man Learn from Aristotle?*; M. Kusch, *Theories of Questions in German-Speaking Philosophy Around the Turn of the Century*; N.-E. Sahlin, *'He is no Good for My Work': On the Philosophical Relations between Ramsey and Wittgenstein.* **II: Formal Tools: Induction, Observation and Identifiability** – T.A.F. Kuipers, *The Carnap-Hintikka Programme in Inductive Logic*; I. Levi, *Caution and Nonmonotic Inference*; I. Niiniluoto, *Inductive Logic, Atomism, and Observational Error*; A. Mutanen, *Theory of Identifiability.* **III: Questions in Inquiry: The Interrogative Model** – S. Bromberger, *Natural Kinds and Questions*; S.A. Kleiner, *The Structure of Inquiry in Developmental Biology*; A. Wiśniewski, *Some Foundational Concepts of Erotetic Semantics*; J. Woleński, *Science and Games.* **IV: Growth of Knowledge: Explanation and Discovery** – M. Sintonen, *Explanation: The Fifth Decade*; E. Weber, *Scientific Explanation and the Interrogative Model of Inquiry*; G. Gebhard, *Scientific Discovery, Induction, and the Multi-Level Character of Scientific Inquiry*; M. Kiikeri, *On the Logical Structure of Learning Models.* **V: Jaakko Hintikka: Replies**.

VOLUME 52 (1997)

Helena Eilstein
LIFE CONTEMPLATIVE, LIFE PRACTICAL.
AN ESSAY ON FATALISM

Preface. **Chapter One: Oldcomb and Newcomb** – 1. *In the King Comb's Chamber of Game*; 2. *The Newcombian Predicaments.* **Chapter Two: Ananke** – 1. *Fatalism: What It Is Not?*; 2. *Fatalism: What Is It?*; 3. *Fatalism and* a priori *Arguments*; 4. *Fatalism and 'Internal' Experience*; 5. *Determinism, Indeterminism and Fatalism*; 6. *Transientism, Eternism and Fatalism*; 7. *Fatalism: What It Does Not Imply?.* **Chapter Three: Fated Freedom** – 1. *More on Libertarianism*; 2. *On the Deterministic Concept of Freedom*; 3. *Moral Self and Responsibility in the Light of Probabilism*; 4. *Fated Freedom.* **Chapter Four: The Virus of Fatalism** – 1. *Fatalism and Problems of Cognition*; 2. *The Virus of Fatalism: Why Mostly Harmless?*

VOLUME 53 (1997)

Dimitri Ginev
A PASSAGE TO THE HERMENEUTIC PHILOSOPHY OF SCIENCE

Preface; **Introduction**: *Topics in the Hermeneutic Philosophy of Science*; **Chapter 0**: *On the Limits of the Rational Reconstruction of Scientific Knowledge*; **Chapter 1**: *On the Hermeneutic Nature of Group Rationality*; **Chapter 2**: *Towards a Hermeneutic Theory of Progressive Change in Scientific Development*; **Chapter 3**: *Beyond Naturalism and Traditionalism*; **Chapter 4**: *A Critical Note on Normative Naturalism*; **Chapter 5**: *Micro and Macrohermeneutics of Science*; *Concluding Remarks.*

VOLUME 54 (1997)

IN ITINERE. EUROPEAN CITIES AND THE BIRTH OF MODERN SCIENTIFIC PHILOSOPHY
(Edited by Roberto Poli)

Introduction – R. Poli, *In itinere. Pictures from Central-European Philosophy.* **Stages of the Tour** – K. Schuhmann, *Philosophy and Art in Munich around the Turn of the Century*; M. Libardi, *In itinere: Vienna 1870-1918*; W. Baumgartner, *Nineteenth-Century Würzburg: The Development of the Scientific Approach to Philosophy*; L. Dappiano, *Cambridge and the Austrian Connection*; J. Sebestik, *Prague Mosaic. Encounters with the Prague Philosophers*; J.J. Jadacki, *Warsaw: The Rise and Decline of Modern Scientific Philosophy in the Capital City of Poland*; J. Woleński, *Lvov*; L. Albertazzi, *Science and the Avant-Garde in Early Nineteenth-Century Florence*; F. Minazzi, *The Presence of Phenomenology in Milan between the Two World Wars. The Contribution of Antonio Banfi and Giulio Preti.*

VOLUME 55 (1997)

REALISM AND QUANTUM PHYSICS
(Edited by Evandro Agazzi)

Introduction – E. Agazzi. **Part One: Philosophical Considerations** – P. Horwich, *Realism and Truth*, E. Agazzi, *On the Criteria for Establishing the Ontological Status of Different Entities*; A. Baltas, *Constraints and Resistance: Stating a Case for Negative Realism*; M. Paty, *Predicate of Existence and Predicability for a Theoretical Object in Physics.* **Part Two: Observability and Hidden Entities** – F. Bonsak, *Atoms: Lessons of a History*; A. Cordero, *Arguing for Hidden Realities*; B. d'Espagnat, *On the Difficulties that Attributing Existence to "Hidden" Quantities May Raise*; M. Pauri, *The Quantum, Space-Time and Observation.* **Part Three: Applications to Quantum Physics** – D. Albert, *On the Phenomenology of Quantum-Mechanical Superpositions*; G.C. Ghiraldi, *Realism and Quantum Mechanics*; M. Crozon, *Experimental Evidence of Quark Structure Inside Hadrons.*

VOLUME 56 (1997)

IDEALIZATION VIII: MODELLING IN PSYCHOLOGY
(Edited by Jerzy Brzeziński, Bodo Krause and Tomasz Maruszewski)

Part I: Philosophical and Methodological Problems of Cognition Process – J. Wane, *Idealizing the Cartesian-Newtonian Paradigm as Reality: The Impact of New-Paradigm Physics on Psychological Theory*; E. Hornowska, *Operationalization of Psychological Magnitudes. Assumptions-Structure-Consequences*; T. Bachmann, *Creating Analogies – On Aspects of the Mapping Process between Knowledge Domains*; H. Schaub, *Modelling Action Regulation.* **Part II: The Structure of Ideal Learning Process** – S. Ohlson, J.J. Jewett, *Ideal Adaptive Agents and the Learning Curve*; B. Krause, *Towards a Theory of Cognitive Learning*; B. Krause, U. Gauger, *Learning and Use of Invariances: Experiments and Network Simulation*; M. Friedrich, *"Reaction Time" in the Neural Network Module ART 1.* **Part III: Control Processes in Memory** – J. Tzelgov, V. Yehene, M. Naveh-Benjamin, *From Memory to Automaticity and Vice Versa: On the Relation between Memory and Automaticity*; H. Hagendorf, S. Fisher, B. Sá, *The Function of Working Memory in Coordination of Mental Transformations*; L. Nowak, *On Common-Sense and (Para-)Idealization*; I. Nowakowa, *On the Problem of Induction. Toward an Idealizational Paraphrase.*

VOLUME 57 (1997)

EUPHONY AND LOGOS
ESSAYS IN HONOUR OF MARIA STEFFEN-BATÓG AND TADEUSZ BATÓG
(Edited by Roman Murawski, Jerzy Pogonowski)

Preface. **Scientific Works of Maria Steffen-Batóg and Tadeusz Batóg** – *List of Publications of Maria Steffen-Batóg*; *List of Publications of Tadeusz Batóg*; J. Pogonowski, *On the Scientific Works of Maria Steffen-Batóg*; Jerzy Pogonowski, *On the Scientific Works of Tadeusz Batóg*; W. Lapis, *How Should Sounds Be Phonemicized?*; P. Nowakowski, *On Applications of Algorithms for Phonetic Transcription in Linguistic Research*; J. Pogonowski, *Tadeusz Batóg's Phonological Systems.* **Mathematical Logic** – W. Buszkowski, *Incomplete Information Systems and Kleene 3-valued Logic*; M. Kandulski, *Categorial Grammars with Structural Rules*; M. Kołowska-Gawiejnowicz, *Labelled Deductive Systems for the Lambek Calculus*; R. Murawski, *Satisfaction Classes – a Survey*; K. Świrydowicz, *A New Approach to Dyadic Deontic Logic and the Normative Consequence Relation*; W. Zielonka, *More about the Axiomatics of the Lambek Calculus.* **Theoretical Linguistics** – J.J. Jadacki, *Troubles with Categorial Interpretation of Natural Language*; M. Karpiński, *Conversational Devices in Human-Computer Communication Using WIMP UI*; W. Maciejewski, *Qualitative Orientation and Gramatical Categories*; Z. Vetulani, *A System of Computer Understanding of Text*; A. Wójcik, *The Formal Development of van Sandt's Presupposition Theory*; W. Zabrocki, *Psychologism in Noam Chomsky's Theory (Tentative Critical Remarks)*; R. Zuber, *Defining Presupposition without Negation.* **Philosophy of Language and Methodology of Sciences** – J. Kmita, *Philosophical Antifundamentalism*; A. Luchowska, *Peirce and Quine: Two Views on Meaning*; S. Wiertlewski, *Method According to Feyerabend*; J. Woleński, *Wittgenstein and Ordinary Language*; K. Zamiara, *Context of Discovery – Context of Justification and the Problem of Psychologism.*

VOLUME 58 (1997)

THE POSTMODERNIST CRITIQUE OF THE PROJECT OF ENLIGHTENMENT
(Edited by Sven-Eric Liedman)

S.-E. Liedman, *Introduction*; R. Wokler, *The Enlightenment Project and its Critics*; M. Benedikt, *Die Gegenwartsbedeutung von Kants aufklärender Akzeptanz und Zurückweisung des Modells der Naturwissenschaft für zwischenmenschliche Verhältnisse: Verfehlte Beziehungen der Geisterwelt Swedenborgs*; S.-E. Liedman, *The Crucial Role of Ethics in Different Types of Enlightenment (Condorcet and Kant)*; S. Dahlstedt, *Forms of the Ineffable: From Kant to Lyotard*; P. Magnus Johansson, *On the Enlightenment in Psycho-Analysis*; E. Lundgren-Gothlin, *Ethics, Feminism and Postmodernism: Seyla Benhabib and Simone de Beauvoir*; E. Kiss, *Gibt es ein Projekt der Aufklärung und wenn ja, wie viele (Aufklärung vor dem Horizont der Postmoderne)*; E. Kennedy, *Enlightenment Anticipations of Postmodernist Epistemology*; L. Nowak, *On Postmodernist Philosophy: An Attempt to Identify its Historical Sense*; M. Castillo, *The Dilemmas of Postmodern Individualism.*

VOLUME 59 (1997)

BEYOND ORIENTALISM. THE WORK OF WILHELM HALBFASS AND
ITS IMPACT ON INDIAN AND CROSS-CULTURAL STUDIES
(Edited by Eli Franco and Karin Preisendanz)

E. Franco, K. Preisendanz, *Introduction and Editorial Essay on Wilhelm Halbfass*; Publications by Wilhelm Halbfass; W. Halbfass, *Research and Reflections: Responses to my*

Respondents. Beyond Orientalism? Reflections on a Current Theme. **Part I: Cross-Cultural Encounter and Dialogue** – F. X. Clooney, SJ, *Wilhelm Halbfass and the Openness of the Comparative Project*; F. Dallmayr, *Exit from Orientalism: Comments on Wilhelm Halbfass*; S. D. Serebriany, *Some Marginal Notes on India and Europe*; R. Sen (née Mookerjee), *Some Reflections on India and Europe: An Essay in Understanding*; K. Karttunen, *Greeks and Indian Wisdom*; D. Killingley, *Mlecchas, Yavanas and Heathens: Interacting Xenologies in Early Nineteenth-Century Calcutta*; W. Halbfass, *Research and Reflections: Responses to my Respondents. Cross-Cultural Encounter and Dialogue.* **Part II: Issues of Comparative Philosophy** – J. Nath Mohanty, *Between Indology and Indian Philosophy;* J. S. O'Leary, *Heidegger and Indian Philosophy*; S. Rao, *"Subordinate" or "Supreme"? The Nature of Reason in India and the West*; R. Ivekoviæ, *The Politics of Comparative Philosophy*; B.-A. Scharfstein, *The Three Philosophical Traditions*; W. Halbfass, *Research and Reflections: Responses to my Respondents. Issues of Comparative Philosophy.* **Part III: Topics in Classical Indian Philosophy** – J. E. M. Houben, *Bhartṛhari's Perspectivism: The Vṛtti and Bhartṛhari's Perspectivism in the First kāṇḍa of the Vākyapadīya*; J. Bronkhorst, *Philosophy and the Vedic Exegesis in the Mīmāṃsā*; J. Taber, *The Significance of Kumārila's Philosophy*; K. Harikai, *Kumārila's Acceptance and Modification of Categories of the Vaiœeṣika School*; V. Lysenko, *The Vaiœeṣika Notions of ākāsa and diœ from the Perspective of Indian Ideas of Space*; B. M. Perry, *Early Nyāya and Hindu Orthodoxy: ānvīkṣikī and adhikāra*; W. Halbfass, *Research and Reflections: Responses to my Respondents: Topics in Classical Indian Philosophy.* **Part IV: Developments and Attitudes in Neo-Hinduism** – A. O. Fort, *Jīvanmukti and Social Service in Advaita and Neo-Vedānta*; S. Elkman, *Religious Plurality and Swami Vivekananda.* **Part V: Indian Religion, Past and Present** – *M. Hara, A Note on dharmasya sūkṣmā gatiḥ*; A. Wezler, *The Story of Aṇī-Māṇḍavya as told in the Mahābhārata: Its significance for Indian Legal and Religious History*; Y. Grinshpon, *Experience and Observation in Traditional and Modern Pātañjala Yoga*; F. J. Korom, *Language Belief and Experience in Bengali Folk Religion*; W. Halbfass, *Research and Reflection: Responses to my Respondents. Developments and Attitudes in Neo-Hinduism; Indian religion, Past and Present.*

VOLUME 60 (1998)

MARX'S THEORIES TODAY

(Edited by Ryszard Panasiuk and Leszek Nowak)

R. Panasiuk, *Introduction*; **Part I: On Dialectics and Ontology** – S.-E. Liedman, *Engels and the Laws of Dialectics*; R. Panasiuk, *On Dialectics in Marxism Again*; R. Albritton, *The Unique Ontology of Capital*; R. Washner, *It is not Singularity that Governs the Nature of Things. The Principle of Isolated Individual and Its Negation in Marx's Doctoral Thesis*; **Part II: On Historical Materialism and Social Theories** – Z. Cackowski, *The Continuing Validity of the Marxian Thought*; P. Casal, *From Unilineal to Universal Historical Materialism*; I. Hunt, *A Dialectical Interpretation and Resurrection of Historical Materialism*; W. Krajewski, *The Triumph of Historical Materialism*; L. Nowak, *The Adaptive Interpretation of Historical Materialism: A Survey. On a Contribution to Polish Analytical Marxism*; M. Kozłowski, *A New Look at Capitalism. Between the* Decommunisation *of Marx's and the* Defeudalisation *of Hegel's Visions of Capitalism*; F. Moseley, *An Empirical Appraisal of Marx's Economic Theory*; Ch. Bertram and A. Carling, *Stumbling into Revolution. Analytical Marxism, Rationality and Collective Action*; K. Graham, *Collectives, Classes and Revolutionary Potential in Marx*; U. Himmelstrand, *How to Become and Remain a Marxicising Sociologist. An Egocentric Report*; **Part III: On Axiology and the Socialist Project** – P. Kamolnick, *Visions of Social Justice in Marx: An Assessment of Recent Debates in Normative Philosophy*; W. Schmied-Kowarzik, *Karl Marx as a Philosopher of Human Emancipation*; H. J. Sandkühler, *Marx – Welche Rationalität? Epistemische Kontexte und Widersprüche der Transformation von Philosophie in Wissenschaft*; J. Kmita, *The*

Production of "Rational Reality" and the "Systemic Coercion"; J. Bidet, *Metastructure and Socialism*; T. Andreani, *Vers une Issue Socialiste à la Crise du Capitalisme*; W. Becker, *The Bankruptsy of Marxism. About the Historical End of a World Philosophy*; D. Aleksandrowicz, *Myth, Eschatology and Social Reality in the Light of Marxist Philosophy.*

VOLUME 61 (1997)

REPRESENTATIONS OF SCIENTIFIC REALITY
CONTEMPORARY FORMAL PHILOSOPY OF SCIENCE IN SPAIN
(Edited by Andoni Ibarra and Thomas Mormann)

Introduction – A. Ibarra, T. Mormann, *The Long and Winding Road to Philosophy of Science in Spain.* **Part 1: Representation and Measurement** – A. Ibarra, T. Mormann, *Theories as Representations*; J. Garrido Garrido, *The Justification of Measurement*; O. Fernandez Prat, D. Quesada, *Spatial Representations and Their Physical Content*; J.A. Díez Calzada, *The Theory-Net of Interval Measurement Theory.* **Part 2: Truth, Rationality, and Method** – J.C. García-Bermejo Ochoa, *Realism and Truth Approximation in Economic Theory*; W.J.Gonzáles, *Rationality in Economics and Scientific Predictions*; J.P. Zamora Bonilla, *An Invitation to Methodonomics.* **Part 3: Logics, Semantics and Theoretical Structures** – J.L.Falguera, *A Basis for a Formal Semantics of Linguistic Formulations of Science*; A. Sobrino, E. Trillas, *Can Fuzzy Logic Help to Pose Some Problems in the Philosophy of Science?*; J. de Lorenzo, *Demonstrative Ways in Mathematical Doing*; M. Casanueva, *Genetics and Fertilization: A Good Marriage*; C.U. Moulines, *The Concept of Universe from a Metatheoretical Point of View.*

VOLUME 62 (1998)

IN THE WORLD OF SIGNS
(Edited by Jacek Juliusz Jadacki and Witold Strawiński)

Introduction. *How to Move in the World of Signs.* **Part I: Theoretical Semiotics** – A. Bogusławski, *Conditionals and Egocentric Mental Predicates*; W. Buszkowski, *On Families of Languages Generated by Categorial Grammar*; K.G. Havas, *Changing the World – Changing the Meaning. On the Meanings of the "Principle of Non-Contradiction"*; H. Hiż, *On Translation*; S. Marcus, *Imprecision, Between Variety and Uniformity: The Conjugate Pairs*; J. Peregrin, P. Sgall, *Meaning and "Propositional Attitudes"*; O.A. Wojtasiewicz, *Some Applications of Metric Space in Theoretical Linguistic.* **Part II: Methodology** – E. Agazzi, *Rationality and Certitude*; I. Bellert, *Human Reasoning and Artifical Intelligence. When Are Computers Dumb in Simulating Human Reasoning?*; T. Bigaj, *Analyticity and Existence in Mathematics*; G.B. Keene, *Taking up the Logical Slack in Natural Languages*; A. Kertész, *Interdisciplinarity and the Myth of Exactness*; J. Srzednicki, *Norm as the Basis of Form*; J.S. Stepanov, *"Cause" in the Light of Semiotics*; J.A. Wojciechowski, *The Development of Knowledge as a Moral Problem.* **Part III: History of Semiotics** – E. Albrecht, *Philosophy of Language, Logic and Semiotics*; G. Deledalle, *A Philosopher's Reply to Questions Concerning Peirce's Theory of Signs*; J. Deledalle-Rhodes, *The Transposition of Linguistic Sign in Peirce's Contributions to "The Nation"*; R. E. Innis, *From Feeling to Mind: A Note on Langer's Notion of Symbolic Projection*; R. Kevelson, *Peirce's Semiotics as Complex Inquiry: Conflicting Methods*; J. Kopania, *The Cartesian Alternative of Philosophical Thinking*; Xiankun Li, *Why Gonsung Long (Kungsun Lung) Said "White Horse Is Not Horse"*; L. Melazzo, *A Report on an Ancient Discussion*; Ding-fu Ni, *Semantic Thoughts of J. Stuart Mill and Chinese Characters*; I. Portis-Winner, *Lotman's Semiosphere: Some Comments;* J. Réthoré, *Another Close Look at the Interpretant*; E. Stankiewicz, *The Semiotic Turn of Breal's "Semantique"*; **Part IV: Linguistic** – K. Heger, *Passive and Other Voices Seen from an Onomasiological Point of*

View; L.I. Komlószi, *The Semiotic System of Events, Intrinsic Temporal and Deictic Tense Relations in Natural Language. On the Conceptualization of Temporal Schemata*; W. M. Osadnik, E. Horodecka, *Polysystem Theory, Translation Theory and Semiotics*; A. Wierzbicka, *THINK – a Universal Human Concept and a Conceptual Primitive*. **Part V: Cultural Semiotics** – G. Bettetini, *Communication as a Videogame*; W. Krysiński, *Joyce, Models, and Semiotics of Passions*; H. Książek-Konicka, *"Visual Thinking" in the Poetry of Julian Przyboœ and Miron Białoszewski*; U. Niklas, *The Space of Metaphor*; M. C. Ruta, *Captivity as Event and Metaphor in Some of Cervantes' Writings*; E. Tarasti, *From Aestetics to Ethics: Semiotics Observations on the Moral Aspects of Art, Especially Music*; L. Tondl, *Is It Justified to Consider the Semiotics of Technological Artefacts?*; V. Voigt, *Poland, Finland and Hungary (A Tuatara's View)*; T.G. Winner, *Czech Poetism: A New View of Poetics Language*; J. Wrede, *Metaphorical Imagery – Ambiguity, Explicitness and Life*. **Part VI: Psycho-Socio-Semiotics** – E.M. Barth, *A Case Study in Empirical Logic and Semiotics. Fundamental Modes of Thought of Nazi Politician Vidkun Quisling, Based on Unpublished Drafts and Notebooks*; P. Bouissac, *Why Do Memmes Die?* W. Kalaga, *Threshold of Signification*, A. Podgórecki, *Do Social Sciences Evaporate?*

VOLUME 63 (1998)

IDEALIZATION IX: IDEALIZATION IN CONTEMPORARY PHYSICS
(Edited by Niall Shanks)

N. Shanks, *Introduction*; M. Bishop, *An Epistemological Role for Thought Experiments*; I. Nowak & L. Nowak, *"Models" and "Experiments" as Homogeneous Families of Notions*; S. French & J. Ladyman, *A Semantic Perspective on Idealization in Quantum Mechanics*; Ch. Liu, *Decoherence and Idealization in Quantum Measurement*; S. Hartmann, *Idealization in Quantum Field Theory*; R. F. Hendry, *Models and Approximations in Quantum Chemistry*; D. Howard, *Astride the Divided Line: Platonism, Empiricism, and Einstein's Epistemological Opportunism*; G. Gale, *Idealization in Cosmology: A Case Study*; A. Maidens, *Idealization, Heuristics and the Principle of Equivalence*; A. Rueger & D. Sharp, *Idealization and Stability:A Perspective from Nonlinear Dynamics*; D. L. Holt & R. G. Holt, *Towards a Very Old Account of Rationality in Experiment: Occult Practices in Chaotic Sonoluminescence.*

VOLUME 64 (1998)

PRAGMATIC IDEALISM. CRITICAL ESSAYS ON NICHOLAS RESCHER'S SYSTEM OF PRAGMATIC IDEALISM
(Edited by Axel Wüstehube and Michael Quante)

Introduction: A. Wüstehube, *Is Systematic Philosophy still Possible?*; T. Airaksinen, *Moral Facts and Objective Values*; L. Rodríguez Duplá, *Values and Reasons*; G. Gale, *Rescher on Evolution and the Intelligibility of Nature*; J Kekes, *The Nature of Philosophy*; P. Machamer, *Individual and Other-Person Morality: A Plea for an Emotional Response to Ethical Problems*; D. Marconi, *Opus Incertum*; M. Marsonet, *Scientific Realism and Pragmatic Idealism*; R. Martin, *Was Spinoza a Person?*; H. Pape, *Brute Facts, Real Minds and the Postulation of Reality: Resher on Idealism and the Ontological Neutrality of Experience*; J. C. Pitt, *Doing Philosophy: Rescher's Normative Methodology*; L. B. Puntel, *Is Truth "Ideal Coherence"?*; M. Quante, *Understanding Conceptual Schemes: Rescher's Quarrel with Davidson*; A. Siitonen, *The Ontology of Facts and Values*; M. Willaschek, *Skeptical Challenge and the Burden of Proof: On Rescher's Critique of Skepticism*; N. Rescher, *Responses.*

VOLUME 65 (1999)

THE TOTALITARIAN PARADIGM AFTER THE END OF COMMUNISM.
TOWARDS A THEORETICAL REASSESSMENT.
(Edited by Achim Siegel)

A. Siegel, **Introduction**: *The Changing Fortunes of the Totalitarian Paradigm in Communist Studies.* **On Recent Controversies Over The Concept Of Totalitarianism** – K. von Beyme, *The Concept of Totalitarianism – A Reassessment after the End of Communist Rule*; K. Mueller, *East European Studies, Neo-Totalitarianism and Social Science Theory*; L. Nowak, *A Conception that is Supposed to Correspond to the Totalitarian Approach to Realsocialism*; E. Nolte, *The Three Versions of the Theory of Totalitarianism and the Significance of the Historical-Genetic Version*; E. Jesse, *The Two Major Instances of Totalitarianism: Observations on the Interconnection between Soviet Communism and National Socialism.* **Classic Concept Of Totalitarianism: Reassessment And Reinterpretation** – J.P. Arnason, *Totalitarianism and Modernity: Franz Borkenau's "Totalitarian Enemy" as a Source of Sociological Theorizing on Totalitarianism*; A. Sölner, *Sigmund Neumann's "Permanent Revolution": A Forgotten Classic of Comparative Research into Modern Dictatorship*; F. Pohlmann, *The "Seeds of Destruction" in Totalitarian Systems. An Interpretation of the Unity in Hannah Arendt's Political Philosophy*; W.J. Patzelt, *Reality Construction under Totalitarianism: An Ethnomethodological Elaboration of Martin Draht's Concept of Totalitarianism*; A. Siegel, *Carl Joachim Friedrich's Concept of Totalitarianism: A Reinterpretation*; M.R. Thompson, *Neither Totalitarian nor Authoritarian: Post-Totalitarianism in Eastern Europe.*

VOLUME 66 (1999)

Leon Gumański

TO BE OR NOT TO BE? IS THAT THE QUESTION?
AND OTHER STUDIES IN ONTOLOGY, EPISTEMOLOGY AND LOGIC

Preface; *The Elements of a Judgment and Existence*; *Traditional Logic and Existential Presuppositions*; *To Be Or Not To Be? Is That The Question?*; *Some Remarks On Definitions; Logische und semantische Antinomien*; *A New Approach to Realistic Epistemology*; *Ausgewählte Probleme der deontischen Logik*; *An Attempt at the Definition of the Biological Concept of Homology*; *Similarity.*

VOLUME 67 (1999)

Kazimierz Twardowski

ON ACTIONS, PRODUCTS AND OTHER TOPICS IN PHILOSOPHY
(Edited by Johannes Brandl and Jan Woleński)

Introduction; Translator's Note; Self-Portrait (1926/91); Biographical Notes. **I. On Mind, Psychology, and Language:** *Psychology vs. Physiology and Philosophy (1897)*; *On the Classification of Mental Phenomena (1898)*; *The Essence of Concepts (1903/24)*; *On Idio- and Allogenetic Theories of Judgment (1907)*; *Actions and Products (1912)*; *The Humanities and Psychology (1912/76)*; *On the Logic of Adjectives (1923/27).* **II. On Truth and Knowledge:** *On So-Called Relative Truths (1900)*; A priori, *or Rational (Deductive) Sciences and* a posteriori, *or Empirical (Inductive) Sciences (1923)*; *Theory of Knowledge. A Lecture Course (1925/75).* **III. On Philosophy:** *Franz Brentano and the History of Philosophy (1895)*; *The Historical Conception of Philosophy (1912)*; *On Clear and Unclear Philosophical Style (1920)*; *Symbolomania and*

Pragmatophobia (1921); Address at the 25th Anniversary Session of the Polish Philosophical Society (1929/31); On the Dignity of the University (1933). **Bibliography**.

VOLUME 68 (2000)

Tadeusz Czeżowski

KNOWLEDGE, SCIENCE AND VALUES. A PROGRAM FOR SCIENTIFIC PHILOSOPHY
(Edited by Leon Gumański)

L. Gumański, *Introduction.* **Part 1: Logic, Methodology and Theory of Science** – *Some Ancient Problems in Modern Form; On the Humanities; On the Method of Analytical Description; On the Problem of Induction; On Discussion and Discussing; On Logical Culture; On Hypotheses; On the Classification of Sentences and Propositional Functions; Proof; On Traditional Distinctions between Definitions; Deictic Definitions; Induction and Reasoning by Analogy; The Classification of Reasonings and its Consequences in the Theory of Science; On the so-called Direct Justification and Self-evidence; On the Unity of Science; Scientific Description.* **Part 2: The World of Human Values and Norms** – *On Happiness; How to Understand "the Meaning of Life" ?; How to Construct the Logic of Goods?; The Meaning and the Value of Life; Conflicts in Ethics; What are Values?; Ethics, Psychology and Logic.* **Part 3: Reality–Knowledge–World** – *Three Attitudes towards the World; On Two Views of the World; A Few Remarks on Rationalism and Empiricism; Identity and the Individual in Its Persistence; Sensory Cognition and Reality; Philosophy at the Crossroads; On Individuals and Existence.* J.J. Jadacki, *Trouble with Ontic Categories or Some Remarks on Tadeusz Czeżowski's Philosophical Views;* W. Mincer, *The Bibliography of Tadeusz Czeżowski.*

VOLUME 69 (2000)

Izabella Nowakowa, Leszek Nowak

THE RICHNESS OF IDEALIZATION

Preface; **Introduction** – *Science as a Caricature of Reality.* **Part I: THREE METHODOLOGICAL REVOLUTIONS** – *1. The First Idealizational Revolution.Galileo's-Newton's Model of Free Fall; 2. The Second Idealizational Revolution. Darwin's Theory of Natural Selection; 3. The Third Idealizational Revolution. Marx's Theory of Reproduction.* **Part II: THE METHOD OF IDEALIZATION** – *4. The Idealizational Approach to Science: A New Survey; 5. On the Concept of Dialectical Correspondence; 6. On Inner Concretization. A Certain Generalization of the Notions of Concretization and Dialectical Correspondence; 7. Concretization in Qualitative Contexts; 8. Law and Theory: Some Expansions; 9. On Multiplicity of Idealization.* **Part III: EXPLANATIONS AND APPLICATIONS** – *10. The Ontology of the Idealizational Theory; 11. Creativity in Theory-building; 12. Discovery and Correspondence; 13. The Problem of Induction. Toward an Idealizational Paraphrase; 14. "Model(s) and "Experiment(s). An Analysis of Two Homogeneous Families of Notions; 15. On Theories, Half-Theories, One-fourth-Theories, etc.; 16. On Explanation and Its Fallacies; 17. Testability and Fuzziness; 18. Constructing the Notion; 19. On Economic Modeling; 20. Ajdukiewicz, Chomsky and the Status of the Theory of Natural Language; 21. Historical Narration; 22. The Rational Legislator.* **Part IV: TRUTH AND IDEALIZATION** – *23. A Notion of Truth for Idealization; 24. "Truth is a System": An Explication; 25. On the Concept of Adequacy of Laws; 26. Approximation and the Two Ideas of Truth; 27. On the Historicity of Knowledge.* **Part V: A GENERALIZATION OF IDEALIZATION** – *28. Abstracts Are Not Our Constructs. The Mental Constructs Are Abstracts; 29. Metaphors and Deformation; 30. Realism, Supra-Realism and Idealization.* **REFERENCES** – *I. Writings on Idealization; II. Other Writings.*

VOLUME 70 (2000)

QUINE. NATURALIZED EPISTEMOLOGY, PERCEPTUAL KNOWLEDGE AND ONTOLOGY
(Edited by Lieven Decock and Leon Horsten)

Introduction. **Naturalized Epistemology** – T. Derksen, *Naturalistic Epistemology, Murder and Suicide? But what about the Promises!*; Ch. Hookway, *Naturalism and Rationality*; M. Gosselin, *Quine's Hypothetical Theory of Language Learning. A Comparison of Different Conceptual Schemes and Their Logic.* **The Nature of Perceptual Knowledge** – J. van Brakel, *Quine and Innate Similarity Spaces*; D. Koppelberg, *Quine and Davidson on the Structure of Empirical Knowledge*; E. Picardi, *Empathy and Charity.* **Ontology** – S. Laugier, *Quine: Indeterminacy, 'Robust Realism', and Truth*; R. Vergauwen, *Quine and Putnam on Conceptual Relativity and Reference: Theft or Honest Toil?*; I. Douven, *Empiricist Semantics and Indeterminism of Reference*; L. Decock, *Domestic Ontology and Ideology*; P. Gochet, *Canonical Notation, Predication and Ontology.*

VOLUME 71 (2000)

LOGIC, PROBABILITY AND SCIENCE
(Edited by Niall Shanks)

N. Shanks & R.R. Gardner, *Introduction*; C. Morgan, *Canonical Models and Probabilistic Semantics (Commentary by François Lepage; Reply by Morgan)*; F. Lepage, *A Many-Valued Probabilistic Logic (Commentary by Charles Morgan; Reply by Lepage)*; P. Rawling, *The Exchange Paradox, Finite Additivity, and the Principle of Dominance (Commentary by Robert R. Gardner; Reply by Rawling)*; S. Vineberg, *The Logical Status of Conditional and its Role in Confirmation (Commentary by Piers Rawling; Reply by Vineberg)*; D. Mayo, *Science, Error Statistics, and Arguing from Error (Commentary by Susan Vineberg; Reply by Mayo)*; M.N. Lance, *The Best is the Enemy of the Good: Bayesian Epistemology as a Case Study in Unhelpful Idealization (Commentary by Leszek Nowak; Reply by Lance)*; R.B. Gardner & M.C. Wooten, *An Application of Bayes' Theorem to Population Genetics (Commentary by Lynne Seymour; Reply by Gardner and Wooten)*; P.D. Johnson, Jr., *Another Look at Group Selection (Commentary by Niall Shanks; Reply by Johnson)*; C.F. Juhl, *Teleosemantics, Kripkenstein and Paradox (Commentary by Daniel Bonevac; Reply by Juhl)*; D. Bonevac, *Constitutive and Epistemic Principles (Commentary by Mark Lance; Reply by Bonevac)*; O. Bueno, *Empiricism, Mathematical Truth and Mathematical Knowledge (Commentary by Chuang Liu; Reply by Bueno)*; Ch. Liu, *Coins and Electrons: A Unified Understanding of Probabilistic Objects (Commentary by Steven French; Reply by Liu)*; A. Maidens, *Are Electrons Vague Objects? (Commentary by David Over; Reply by Maidens).*

VOLUME 72 (2000)

ON COMPARING AND EVALUATING SCIENTIFIC THEORIES
(Edited by Adam Jonkisz and Leon Koj)

L. Koj, *Preface*; L. Koj, *Methodology and Values*; L. Koj, *Science as System*; A. Grobler, *Explanation and Epistemic Virtue*; P. Giza, *"Intelligent" Computer System and Theory Comparison*; H. Ogryzko-Wiewiórowski, *Methods of Social Choice of Scientific Theories*; K. Jodkowski, *Is the Causal Theory of Reference a Remedy for Ontological Incommensurability?*; W. Balzer, *On Approximative Reduction*; C. Ulises Moulines, *Is There Genuinely Scientific Progress?*; A. Jonkisz, *On Relative Progress in Science.*

VOLUME 73 (2000)

THE RATIONALITY OF THEISM
(Edited by Adolfo García de la Sienra)

Preface; A. García de la Sienra, *Introduction*; W. Redmond, *A Logic of Religious Faith and Development*; J.M. Bocheński, O.P., *The Five Ways*; M. Beuchot, *Saint Thomas' Third Way: Possibility and Necessity, Essence and Existence*; R. Swinburne, *Cosmological and Teleological Arguments*; A. García de la Sienra, *The Ontological Argument*; A. García de la Sienra, *Pascal's Wager*; A. García de la Sienra & A. Araujo, *The Experience of God in Moral Obligation*; A. Plantinga, *The onus probandi of Theism*; R. A. Clouser, *Is God Eternal?*; A. Tomasini, *The Presence and Absence of God*; L. Nowak, *On the Common Structure of Science and Religion.*

VOLUME 74 (2000)

POLISH PHILOSOPHERS OF SCIENCE AND NATURE IN THE 20th CENTURY
(Edited by Władysław Krajewski)

W. Krajewski, *Introduction*; **I. Philosophers** J. Woleński, *Tadeusz Kotarbiński – Reism and Science*; A. Jedynak, *Kaziemierz Ajdukiewicz – From Radical Conventionalism to Radical Empiricism*; L. Gumański, *Tadeusz Czeżowski – Our Knowledge though Uncertain is Probable*; M. Tałasiewicz, *Jan Łukasiewicz – The Quest for the Form of Science*; I. Szumilewicz-Lachman, *Zygmunt Zawirski – The Notion of Time*; A. Jedynak, *Janina Hosiasson-Lindenbaumowa – The Logic of Induction*; T. Bigaj, *Joachim Metallmann – Causality, Determinism and Science*; J. Woleński, *Izydora Dąmbska – Between Conventionalism and Realism*; A. Koterski, *Henryk Mehlberg – The Reach of Science*; I. Nowakowa, *Adam Wiegner's Nonstandard Empiricism*; W. Krajewski, *Janina Kotarbińska – Logical Methodology and Semantic*; M. Tałasiewicz, *Maria Kokoszyńska-Lutmanowa – Methodology, Semantics, Truth*; T. Batóg, *Seweryna Łuszczewska-Romahnowa – Logic and Philosophy of Science*; M. Omyła, *Roman Suszko – From Diachronic Logic to Non-Fregean Logic*; J. Woleński, *Klemens Szaniawski – Rationality and Statistical Methods*; A. Jedynak, *Halina Mortimer – The Logic of Induction*; K. Zamiara, *Jerzy Giedymin – From the Logic of Science to the Theoretical History of Science*; J. M. Dołęga, *B. J. Gawecki – A Philosopher of the Natural Sciences*; A. Bronk, *Stanisław Kamiński – A Philosopher and Historian of Science*; Z. Hajduk, *Stanisław Mazierski – A Theorist of Natural Lawfulness.* **II. Scientists** W. Krajewski, *Marian Smoluchowski – A Forerunner of the Chaos Theory*; A. Motycka, *Czesław Białobrzeski's Conception of Science*; M Tempczyk, *Leopold Infeld – The Problem of Matter and Field*; M. Czarnocka, *Grzegorz Białkowski – Science and Its Subject*; J. Płazowski, *Jerzy Rayski – Physicist and Philosopher of Physics*; J. Misiek, *Zygmunt Chyliński – Physics, Philosophy, Music*; W. Sady, *Ludwik Fleck – Thought Collectives and Thought Styles.* **II. General Surveys** K. Ajdukiewicz, *Logicist Anti-Irrationalism in Poland*; K. Szaniawski, *Philosophy of Science in Poland*; I. Nowakowa, *Main Orientations in the Contemporary Polish Philosophy of Science.*

VOLUME 75 (2000)

STRUCTURALIST KNOWLEDGE REPRESENTATION
PARADIGMATIC EXAMPLES
(Edited by Wolfgang Balzer, Joseph D. Sneed and C. Ulises Moulines)

W. Balzer, U. Moulines, *Introduction*; J. A Diez Calzada, *Structuralist Analysis of Theories of Fundamental Measurement*; A. García de la Sienra, P. Reyes, *The Theory of Finite Games in Extensive Form*; H. J. Burscheid, H. Struve, *The Theory of Stochastic Fairness – its Historical Development, Formulation and Justification*; W. Balzer, R. Mattessich, *Formalizing the Basis of Accounting*; W. Diederich, *A Reconstruction of Marxian Economics*; B. Hamminga, W. Balzer, *The Basic Structure of Neoclassical General Equilibrium Theory*; K. Manhart, *Balance Theories: Two Reconstructions and the Problem of Intended Applications*; R. Westermann, *Festinger's Theory of Cognitive Dissonance: A Structuralist Theory-Net*; R. Reisenzein, *Wundt's Three-Dimensional Theory of Emotion*; P. Lorenzano, *Classical Genetics and the Theory-Net of Genetics*; H. Hettema, T.A.F. Kuipers, *The Formalisation of the Periodic Table*; C. Ulises Moulines, *The Basic Core of Simple Equilibrium Thermodynamics*; T. Bartelborth, *An Axiomatization of Classical Electrodynamics*; *Author's Index*; *Subject Index.*

VOLUME 76 (2000)

EVENTS, FACTS AND THINGS
(Edited by J. Faye, U. Scheffler and Max Urchs)

J. Faye, U. Scheffler and M. Urchs, *Philosophical Entities: An Introduction*; J. Faye, *Facts as Truth Makers*; J. Persson, *Examining the Facts*; U. Scheffler and Y. Shramko, *The Logical Ontology of Negative Facts: On What is Not*; W. Stelzner, *The Impact of Negative Fact for the Imaginary Logic of N.A. Vasil'ev*; B. Rode Meinertsen, *Events, Facts and Causation*; U. Meixner, *Essential Conception of Events*; P. Stekeler-Weithofer, *Questions and Theses Concerning (Mental) Events and Causation*; E. Tegtmeier, *Events as Facts*; M. Urchs, *Events of Episystems*; J. Seibt, *The Dynamic Constitution of Things*; K. Trettin, *Tropes and Things*; D. von Wachter, *A World of Fields*; A. Bartels, *Quantum Field Theory: A Case for Event Ontologies?*; M. Dorato, *Facts, Events, Things and the Ontology of Physics*; M. Kuhlmann, *Processes as Objects of Quantum Field Theory: Consequences for the Interpretation of QFT*; J. Paśniczek, *Objects vs. Situations*; A. Siitonen, *Effects or Consequences of Action*; P. Needham, *Hot Stuff*; L. Bo Gundersen, *Goodman's Gruesome Modal Fallacy*; Th. Mormann, *Topological Representations of Mereological Systems*; U. Scheffer & M. Winkler, *Tools.*